ALGEBRA.ZIP

BASIC ALGEBRA I

Version 2A

Table of Contents

Chapter 1: Equations, Relationships, and Functions

1.1 Equations and Expressions	1
Vocabulary and Exercise	7
1.2 Relationships and Linear Functions	10
Vocabulary and Exercise	23
1.3 System of Equations with Two Variables	25
Vocabulary and Exercise	31
Chapter 1 Summary	33
Chapter Exercise Answer Key	36

Chapter 2: Polynomials and Quadratics

2.1 Operations with Polynomials	39
Vocabulary and Exercise	45
2.2 Quadratic Equations	46
Vocabulary and Exercise	53
2.3 Solving Quadratics	55
Vocabulary and Exercise	59
2.4* Expansion of Quadratics and Parabolas	60
Chapter 2 Summary	67
Chapter Exercise Answer Key	69

Chapter 3: System of Equations and Functions

3.1 System of Equations with Three Variables	72

Vocabulary and Exercise	*76*
3.2 Systems of Equations with None-Linear Equations	*78*
Vocabulary and Exercise	*84*
3.3 Piecewise Functions and Discontinuity	*86*
Vocabulary and Exercise	*92*
Chapter 3 Summary	*94*
Chapter Exercise Answer Key	*97*

Symbols and Definition

Symbols	Definition	Example
$\mathbb{N}$	All natural numbers	1,2,3,4,5,6,7,8,…
$\mathbb{R}$	All real numbers	1,2,3,4,5,6,7,8, 3.33333…
$\cup$	Union	$(1, 5) \cup (5, +\infty)$ 1 ~ 5(not included), 5 ~ ∞(not included)
∞	Infinity	-
$\forall$	For all	$\forall x \geq 0 : \sqrt{x}$ is defined
$\emptyset$	Null (empty set)	-
$\exists$	Exists	$\exists x \geq 0 : \sqrt{x} = 3$
$\subset$	A subset of…	A = {1, 3, 5, 7, 9} $A \subset \mathbb{N}$
$\not\subset$	Not a subset of…	A = {$\sqrt{-1}$} $A \not\subset \mathbb{N}$
$\in$	An element of …	$x = 10$ $x \in \mathbb{N}$
$\notin$	Not an element of…	$x = \sqrt{-1}$ $x \notin \mathbb{N}$
:=	Definition	$f(x) := x^2$

| ⇔ | A theorem that if A is true then B is also true | $f(x) := x^2$ $f(x) = 0 \Leftrightarrow x = 0$ |

Welcome!

Math, especially algebra, is the most fascinating will give its students much satisfaction. As what you read from the content outline page, this book is everything about algebra. From the most basic expressions and functions to derivatives and differentiability. Sure enough, this will be the key for your math journey.

In fact, this book is the very much the bare outline of topics that is chosen. Since the purpose of this is to encourage you to think and explore by yourself. Math is not the subject like English where majority of it will be your teacher standing in front of the class and talk. What math prepares people is the way of thinking and problem solving. Therefore, use your logic and thinking skills, let them aid you along the way of exploring the world of math.

Use Internet and all resources available to you. The answer keys and summaries given at the end of each chapter are to help your understanding of this chapter. However, this does not mean to overload yourself with all information in this book.

You do not need to finish all section assessment questions!

Do the ones that what you believe is useful to yourself. Some of the information that covered in this book is not needed for current stage. However, such were added on the idea of expanding student's knowledge.

Due to the lack of knowledge and attention to details of the author, there may be errors and issues within this textbook. The author asks both the students and teachers who used this book to supply any suggestions. All the suggestion will be appreciated. Please send an email having your suggestions to haimo.li@icloud.com.

The Author
December 16, 2022

Version 2A Preface

Through the publications of the first version, several issues including grammatic errors and symbol usages are to be found as errors. Therefore, in this version, such issues has been removed. More illisutrations and better formatting has been applied. I hope that this version can improve your experience in learning algebra. Sections that are optional regarding personal ability are marked by "*".

The Author
January 17, 2023

Chapter 1 Equations, Relationship and Functions

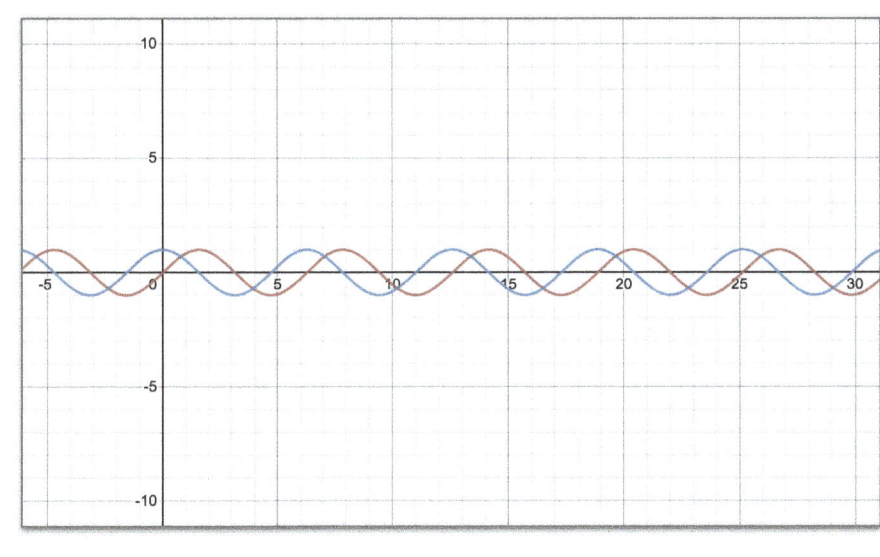

The most basic trignometry functions: sine and cosine. Their graphs are similar —— both are waves.
In this chapter, we will be discovering the most basic relationships —— functions and the methods of graphing such functions.

1.1 Equations and Expressions

Example 1 Identify the following types of mathematical expressions:
- a. $x = 10$
- b. $y \neq 30$
- c. $10x$
- d. $50x + 80y(\pi r^2)$
- e. 2.8×10^2

To identify the types of any mathematical expression, it is critical that you master their specific characteristics. In the sample question above. Three types of expressions are specified. They are equation, inequality, monomial, and polynomial. These are the most fundamental types of equations in math.

Definition 1-1

Variables (letters) and numbers that are joined by multiplication or division are called **monomials**. Thus, one number or variable is also called a monomial.

Definition 1-2

An **equation** expresses a certain relationship where two separate monomials are connected by the equal symbol (=).

Definition 1-3

Inequality is a certain relationship where two monomials are *not* joined by an equal symbol.

Definition 1-4

A **polynomial** is two or more monomials that are joined by addition or subtraction.

SOLUTION

a. Since the given, $x = 10$, both sides of the equal sign are valid monomials. This is identified as an equation.

b. If the given is not joined by an equal sign, then the given is an example of an inequality.

c. Within given, there is no sign of operation or equal sign. Thus, this equation is considered a monomial. Since both expressions are joined by multiplication.

d. In $50x + 80y(\pi r^2)$, we can identify two monomials ($50x$ and $80y(\pi r^2)$). However, pay close attention to how these two monomials are joined. Since they are joined by addition, this is a polynomial.

e. 2.8×10^2 can be evaluated into 280. By definition of a monomial, this is considered a monomial.

Exercise 1 Identify the following expressions.

a. x
b. $y + z$
c. $a \neq b$
d. $g = 10$

As what had mentioned, a polynomial is several monomials joined by additions and subtractions. From what you have previously learned, to evaluate something meaning to calculate the value of an expression. For example, $25 + 8 = 33$. Similarly, a polynomial can be evaluated. Before evaluating a polynomial, you must learn what each polynomial consists of.

Example 2 Identify the degree, constant, and coefficient of the polynomial $x^2 + 6x + 5$.

In this example, we are given the polynomial $x^2 + 6x + 5$ and asked to identify the degree, constant, and coefficient of it. Each of their definitions is given as:

Definition 1-5
Each monomial in a polynomial is called a **term**.

Definition 1-6
The greatest power of any term is the **degree** of the polynomial.

Definition 1-7
Any term that only consists of a number is called a **constant**.

Definition 1-8
All numbers multiplied by a variable are called **coefficients**.

SOLUTION
Thus, in the polynomial $x^2 + 6x + 5$, each of the terms are: x^2, $6x$, and 5. The third term is only consisted of 5, so it is the constant. The greatest power in this polynomial is 2 in x^2, so the degree is 2. In the middle term, 6 is multiplied by x; thus, the coefficient is 6. However, x^2 is multiplied by 1. So, the coefficient is 1 and 6.

Exercise 2 Identify the degree, constant, and coefficient of the polynomial $x^3 + x^2 - x + 5$.

To evaluate a polynomial like the ones that appeared in example 2, we have to know the value of variables. Thus, a sample question may look similar to: *Evaluate the polynomial $x^2 + 6x + 5$ where $x = 3$*. The key to solving this type of problem is to plug in numbers and carefully evaluate the equation.

Example 3 Evaluate the polynomial $x^2 + 6x + 5$ where $x = 3$.

Key: substitute all x with 3. Then carefully evaluate the polynomial.

SOLUTION

Given $x^2 + 6x + 5$ and $x = 3$, by substituting x with 3, we obtain:

$3^2 + 6(3) + 5$ *Substitute*
$= 9 + 18 + 5$ *Evaluate $3^2 = 9$, $6(3) = 18$*
$= 27 + 5$
$= 32$ *Simplify*

Example 4   Evaluate the polynomial $x^2 + 6x^2 + 17x + 3x$ where $x = 2$.

In this example, we encountered several terms that were constructed similarly. For example, the first and second terms both consisted of a x^2. Recall what you have learned previously in math, by using parentheses you can give priority to a series of operations. So, the polynomial can be rewritten as: $(x^2 + 6x^2) + (17x + 3x)$. Now look closer within the parentheses, what similarities have you spotted? Within the first set, both terms are consisted with x^2. Thus, these two terms are called **like terms**.

Definition 1-9
Terms in polynomials have the same variable and power are called **like terms**.

We can combine the like terms into one single term by the foil method. The polynomial can be written as: $x^2(1 + 6) + x(17 + 3)$. By evaluating the inside of the parentheses, we obtain:
$7x^2 + 20x$. The process of using the foil method to rewrite the equation is called **combining like terms**.

SOLUTION

Given the polynomial $x^2 + 6x^2 + 17x + 3x$, and $x = 2$:

Original $= x^2(1 + 6) + x(17 + 3)$
$= 7x^2 + 20x$ *Combine like terms*
$= 7(2^2) + 20(2)$ *Evaluate when $x = 2$*
$= 7(4) + 40$
$= 28 + 40$

$$= 68 \qquad \textit{Simplify}$$

Exercise 3 Evaluate the following polynomials:
    a. $x^4 + 2x^4 + 7x + 8x$ when $x = 3$
    b. $x^2 + 17x + 3$ when $x = 5$

Recall the definition of an equation. It demonstrated a certain relationship. An equation will always consist of one or more variables. The process of finding the value of the variable is called solving the equation. When solving equations, properties of algebra can be used. However, all of the process must be following the property of an equation.

Property 1-1   You may add or subtract the same number on both sides of the equation.

Property 1-2   You may multiply and divide both sides by the same number except for 0.

Example 5 Determine which property is violated.
    Given $5x + 3 = 13$:
$$5x + 3 - 3 = 13 - 3$$
$$5x = 10$$
$$5x \div 0 = 10 \div 0$$
$$0 = 0$$

SOLUTION

From what is demonstrated, the second law is violated. By the definition of solving an equation, and the property of the equation, you should not divide both sides by 0. This is an undefined operation. Secondly, the aim of solving an equation is to find out what the variable's value is. However, in this case, the variable is eliminated.

From what is demonstrated incorrectly in example 5, we can infer what we should do when solving for an equation. This all can be concentrated into 3 key points: 1. Isolate the variable containing the term(s); 2. Combine like term(s); 3. Solve for the variable.

Example 6   Solve the equation: $5x + 3 = 13$.

Key: follow the 3 key points of solving an equation, isolate, combine, and solve.

SOLUTION

Given the equation $5x + 3 = 13$:

$5x + 3 - 3 = 13 - 3$ *Isolate x term by subtracting 3*
$5x = 10$ *Combine like terms*
$5x \div 5 = 10 \div 5$ *Isolate x by dividing 5*
$x = 2$ *Simplify*

Solving for the first time may be difficult for many students. However, there is no short cuts for solving equations. Practice more with equations to directly see like terms and strategies to isolate the variable term.

Exercise 4   Solve: $30x + 5 = 95$

Equations are important in real-life situations. When using equations to solve word problems, the key is to master the relationship between the scenarios. By assuming the answer to the question as known, it is easy to write out the equation. Consider the example:

Example 7   A factory which produces soda cans reports that each can costs x dollars to manufacture and there is a fixed cost of $200. If they manufacture 300 cans, their cost will be $600. What is the cost when the factory produces 600 cans?

Key: there is a fixed relationship between the costs of cans and the number they produce. By assuming the x dollar cost, we are able to write the equation and solve for x, then evaluate the polynomial.

SOLUTION

Given each can costs x dollars to manufacture, the fix cost of $200, and assume that there are n cans being manufactured. The total cost c:

$c = nx + 200$	*write-out relationship.*
$600 = 300x + 200$	*solve for x*
$300x = 400$	
$x = \frac{4}{3}$	
Thus, $c = \frac{4}{3}x + 200$	*write out full relationship*
$c = \frac{4}{3}(600) + 200$	*substitute variables*
$c = 800 + 200$	
$c = 1000$	*simplify*

Exercise 5 A bike company designed a bike that will cost x dollars to manufacture. With the fix cost of $3,000, manufactured 10 bikes will cost $5,000. What is the cost of manufacturing 100 bikes?

1.1 Vocabulary and Exercise

A. Fill in the blank with proper terms.

1. A _____ is a combination of variable and numbers which joined by multiplication.

2. An equation is joined by _____ symbol.

3. Expression $2x > 3$ is a _____.

4. Give an example of a polynomial: _____.

5. For the polynomial $5x^2 + 2x + 3$, identify the following:
   a. The degree of this polynomial is _____.
   b. Identify all terms: _____.
   c. The constant is _____.
   d. All coefficients are: _____.

6. Write two monomials that are like terms: _____, _____.

7. Combine the like terms: $2x^5 + 4x^4 - 3x^3 + (x^2 - x + 3) + 3x^5 - 4x^3$

B. Determine type of the following expressions.
1. 5
2. 30
3. $6y$
4. $7a$
5. $8x^2$
6. $9x^3$
7. $3x + 6x$
8. $3y + 6$
9. $5a = 7$
10. $6x \neq 8$

C. Determine the degree, coefficient and constant
1. $5x + 3$
2. $9x + 6$
3. $x^2 + x + 7$
4. $x^4 + 8x - 8$
5. $x(6x + 3) - 4$
6. $y(2x + 1) - 5$
7. $n(8x + 2z) + y(x + 7)$
8. $z(z + x) + x(y + z)$
9. $7x(5x + 3y)$
10. $9y(55 + 47k)$

D. Evaluate or simplify the polynomials when $x = 6$

1. $5x + 6$
2. $x^2 - 9$
3. $88 - x$
4. $x(x - 9)$
5. $m(y - x)$
6. $5x(8k - 8)$
7. $5x^2(x - 6)$
8. $2[x(x^2)]$
9. $11^2[3x(4x^2 - 1) - 6x]$
10. $4 - y(x^4)$

E. Solve the following equations
1. $4(2x - 5) = 6$
2. $12x - 6 = 30$
3. $5x - 6 = 26$
4. $6x + 77 = 177$
5. $9r - 7 = -1$
6. $3x - 8 = 5x + 2$
7. $k = -4k$
8. $x - 6x + 3 = 0$
9. $n - 5 = 5$
10. $x - 4x(5 - 4) = -1$

F. Solve the following word problems

1. Given that there are 400 legs in a cage, if a chicken only has 2 legs, how many chickens are there?

2. One pencil cost $0.5; one pays $30 for a pencil only. How many pencils are purchased?

3. There is a discount on scripts of a play. Each copy of a script is $20. If one purchases more than 300 scripts, they may get a 30% discount. If in total of $3000 is used to purchase scripts, how many scripts was purchased?

1.2 Relationship and Function

All equations demonstrated a certain relationship. There are several relationships that an equation can represent.

Definition 1-10
One-to-one relationship: the simplest relationship where one x will only have one corresponding value.

Definition 1-11
One-to-more relationship: the relationship where one x may have more than one corresponding value.

Definition 1-12
More-to-one relationship: the relationship where several variables may have one corresponding value.

An equation that we encountered in the previous example all has a one-to-one relationship. These equations are called functions. A typical analogy of imainging the relationship as shareing a tissue with others. Just imagine that your classmates are sick with a running nose. To be a great person, you decided to share a tissue with them. An one to one relationship will let everybody to share one tissue with every sick person. One person share the tissue to one person and when he used that tissue, you give that tissue to another person. Disgusting, isn't it? Therefore, this is not a function. Similarly, a more to one relationship will allow multiple person share the tissue with one sick person.

Generally, the example will show the property of relationships. Only in the case which all person is happy, therefore the relationship is a function.

Definition 1-13A
An equation where one input has one and one only output is called a **function**.

Mathematically, a relationship is a **mapping**. A function is a specific type of mapping where only one input has exactly one output. A mapping is defined as:

Definition 1-14
With two non-empty set A and B, there is one relationship f that for every element in A there is always one and one only element in B that corresponding with that. This relationship is written as $f: A \to B$. Each element is written as
$$b = f(a).$$

You may recognize the definition of mapping to be similar to the definition of a function. In fact, all functions are a mapping of two sets. Thus, a definition of a function can be rewritten in terms of mapping as:

Definition 1-13B
With two non-empty sets A and B where $A \subset \mathbb{R}$ and $B \subset \mathbb{R}$. Given a method f where $f: A \to B$. Therefore, every element in A has one and one only corresponding element in B. If f satisfied this definition, f is a function.

In order to simplify this definition, we introduce logic and math symbols including: $\subset$, $\forall$, and $\exists$. These symbols will be widely used in the following chapters, so consider writing them down.

Definition 1-15

"⊂" means A is a *subset* of B. See the definition above where A ⊂ ℝ, this means that A is a subset of ℝ which represented all real numbers.

Definition 1-16

The symbol "∀" represents *all* the expressions after the colon will *always be true* when the first expression is valid. Such that ∀x ⊂ ℕ: $x > 0$.

Definition 1-17

"∃" is defined as if there exists. The expression after the colon will always be true when the expression in front is valid. For example: ∃x ⊂ ℕ: x is even.

With these logic symbols, we are able to obtain the definition of function as:
D ⊂ ℝ and R ⊂ ℝ,
∃f which satisfies $f: D \to R$.

The definition implies that a function should have either a one-to-one relationship or a more-to-one relationship. Essentially, remember that a function's input and output are an immutable set called an **ordered pair** where one input can have only one output.

The definition of a function can be visual in to the following graphics:

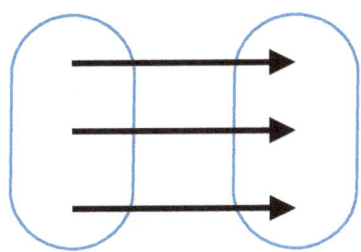

One to One: Function

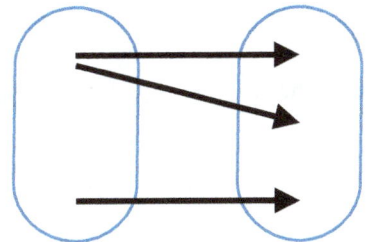

One to more: NOT function

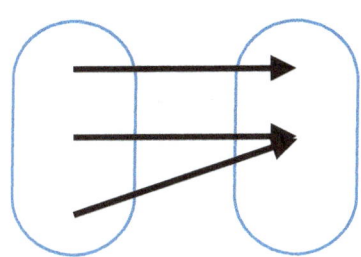

More to One: Function

Example 1 Determine the following relationship.

x	y
1	1
2	5
3	7
4	9

Key: See if one x has more than one corresponding y value or more than one x have the same y.

Solution: The relationship shown in the table is a one-to-one relationship where one *x* has only one *y*.

As what is demonstrated in the example, the relationship has the potential of being a function. However, any function will have a constant rate of change. Where the increase or decrease of *x* and *y* are constant values. This value is called a **slope**.

Definition 1-18
The **slope** is the rate of change in a function. Represented by letter *m*.

Definition 1-19
Slope will always be equal to the change in y-value divided by the change in x-value.

Example 2 Find the slope of the relationship.

x	y
1	2
2	4
3	6
4	8

SOLUTION

The slope equals the change in y divided by the change in x; therefore, 2/1 = 2. The slope is 2.

Exercise 1 Find the slope of the given relationship.

x	y
1	3
2	6
3	9
4	12

A function is usually represented by the letters f, g, and h. When writing a function, we often write the function letter, and the **independent variable** in parentheses.

Definition 1-20
A **dependent variable** is the variable that is the result of the operation(output).

Definition 1-21
An **independent variable** is the input of the function.

When writing a function, we often write as: $f(x) = \ldots$. This format will be read as: "f of x is equal to …". After the equal sign, there is the expression of the main function. For example,
$f(x) = 2x + 1$ is read as "f of x is equal to two x plus one."
Since a function represents a relationship, if we graph the function, we are able to see the direct relationship.

Example 3 Graph the function $f(x) = x + 1$.

This example asked us to graph this function. In order to graph a function, we must know the basic form of a function.

Definition 1-22
The **y-intercept** is the point where the graph of the function intersects with the y-axis.

Definition 1-23
A **Slope Intercept Form** is often expressed as $f(x) = mx + b$. In this format, m is the slope and b are the y-intercept.

With this information, refer back to the function.

Key: identify the slope and y-intercept and graph the equation.

SOLUTION 1

Given the function $f(x) = x + 1$. The y-intercept is at point (0, 1) and the slope m is 1. Then make a x-y table: calculate the value of the function by substituting x with numbers in the table. Then plot the points in a coordinate plane and connect all points.

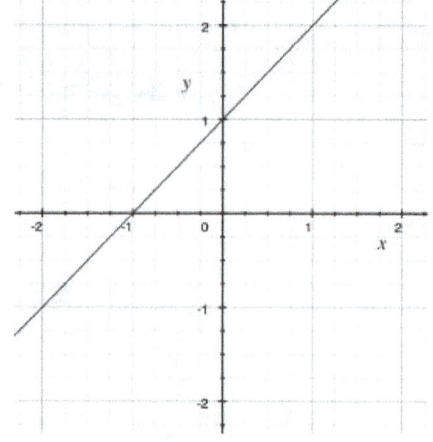

x	y
−2	−1
−1	0
0	1
1	2
2	3

SOLUTION 2

Given the function $f(x) = x + 1$, where the y-intercept is at 1 and slope is also 1. Since slope is the change in y divided by the change in x, we can write the slope in the form of a function: $\frac{1}{1}$. The numerator is the change in *y* and the denominator is the change in *x*. The change indicated how many units does the point go from the y-intercept. Thus, we can have points (0, 1), (1, 2). Plot those points and connect them.

It is important to master the method, the way of graphing and finding slopes of lines. It is difficult when exploring calculus without knowing the definition of slope.

Exercise 2   Graph the function $f(x) = 2x - 3$

We have discussed the method of graphing a function and the basic form of a function: $f(x) = \ldots$.
This proved that a line can be modeled into an expression. When writing a line informing an equation,
we have to introduce another variable *y*. Consider the following example:

Example 4   Write the equation in point slope form that satisfy the line.

In geometry, we define a line as a set of points. An algebraic expression or function is able to represent the relationship of this set of points.

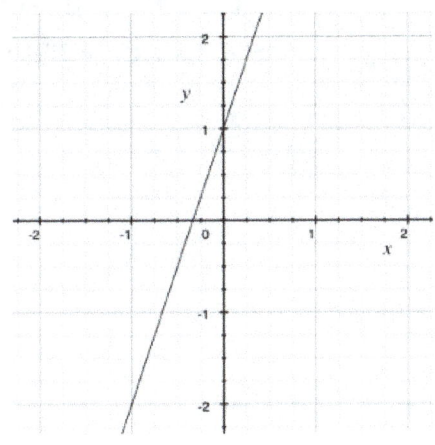

Definition 1-24

A **Point Slope Form** is the form that uses a point on the line and the slope to define the line.

Definition 1-25

A point slope form is written as: $y - y_1 = m(x - x_1)$

Key: find the y-intercept and another point at which the line passes. With that, find the slope of the line.

SOLUTION

Given the graph, the y-intercept is (0, 1) and the line also passes through points.
(–1, –2) with these points:

$$m = (1 + 2)/(0 + 1) \quad \textit{Use the formula for m}$$
$$= 3/1$$
$$= 3 \quad \textit{Simplify}$$

With $m = 3$:
$$y + 2 = 3(x + 1)$$

Example 5 With the function found in example 4, write it in general and slope intercept form.

Example 5 asked us to write the function in general form. A **general form** is another variant of a form of a linear equation.

Definition 1-26
The general form is $a + by = c$

Key: with the given form in point slope form, use algebra to solve for the general and slope intercept form.

SOLUTION

Given the equation $y + 2 = 3(x + 1)$:

$$\text{Original} = y + 2 = 3x + 3 \quad \textit{Distribute 3}$$
$$y + 2 - 2 = 3x + 3 - 2 \quad \textit{Subtract 2}$$
$$y = 3x + 1$$
$$y - 3x = 1 \quad \textit{Simplify}$$

Exercise 3 Write the equation in slope-intercept form that satisfy the graph and rewrite the equation in general form as well as point slope form.

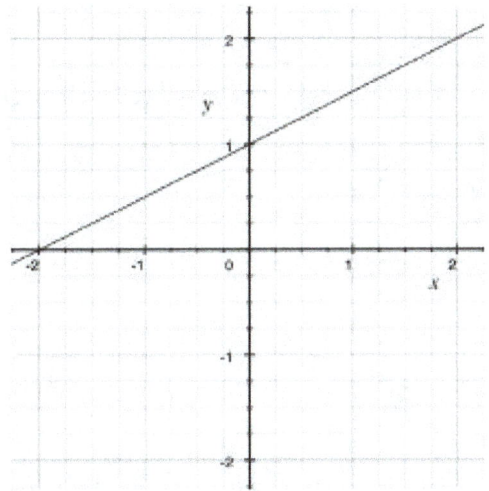

All functions will have a **parent function** which is simply the most basic form of a function. A parent function's expression is a monomial. The functions that we have been discussing for the past two lessons are all linear where all of their graph is a straight line. Thus, we are able to write the parent function of a linear equation (function) as $f(x) = x$. The graph of that function is shown on the left.

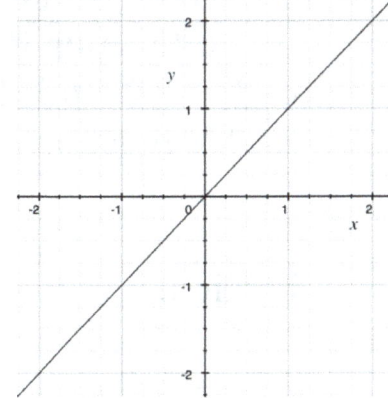

Based on the parent function, we are able to shift the graph by adding or subtracting constants to it. This process is called **transformation**.

Definition 1-27
$f(x) = x + a$ shift the function **up** by a units

Definition 1-28
$f(x) = x - a$ shift the function **down** by a units

Definition 1-29

$f(x) = ax; a \neq 0$ when a > 0, this will **increase the slope** of the function where the line will get steeper. When a < 0, this will **flip the graph**. This flipped graph is symmetrical to the original graph by y-axis.

Example 6 Tell how the function is shifted regarding the parent function.
a. $f(x) = x + 4$
b. $f(x) = x - 2$
c. $f(x) = -3x$
d. $f(x) = 2x$

SOLUTION
a. $f(x) = x + 4$ This function satisfies scenario 1 — $f(x) = x + a$. Thus, it is shifted up 4 units.
b. $f(x) = x - 2$ This function satisfies scenario 2 — $f(x) = x - a$. Thus, this function is shifted down 2 units.
c. $f(x) = -3x$ this function satisfies scenario 3 — $f(x) = ax$. Since $-3 < 0$, the function is reflected by the y-axis and with the slope of -3.
d. $f(x) = 2x$ this function satisfies scenario 3 — $f(x) = ax$. Since $2 > 0$, this function is "steeper" than the parent function with the slope of 2.

Exercise 4 Tell how $f(x) = -5x + 12$ is shifted regarding the parent function.

___

Recall the definition of a function. We have discussed how to determine if a relationship is a function algebraically. However, it is easy to determine that graphically. To implement that, we are using a method called the **vertical line test**.

Definition 1-30
A **vertical line test** is the method which tests whether the given relation is a function or not.

Example 7 Use a vertical line test to determine if the graph given is a function.

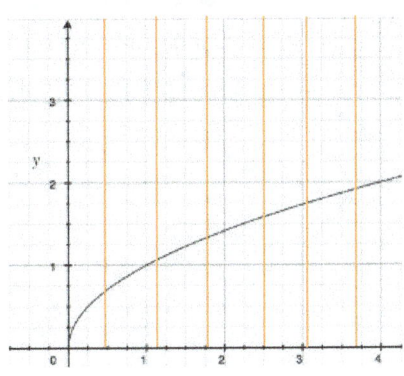

The example asked us to use the vertical line test to determine whether the graph on the right is a function or not. To perform a vertical line test, you will have several vertical lines. See if any of the lines touch the graph more than once. If so, it is not a function.

Key: draw several vertical lines on the graph to the right and see if it passes the vertical line test.

SOLUTION
Since none of the vertical lines passes through the graph more than once, the graph shown on the right is a function.

Exercise 5 Determine whether the graph on the right is a function or not.

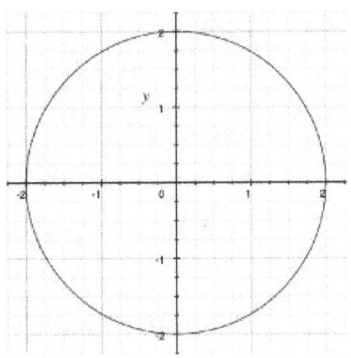

In the previous lessons, we mentioned a vertical line. So, does a vertical line pass the vertical line test? The answer is simple: no. However, what type of expression will be graphed like a vertical line? And what will be graphed as a horizontal line?

Example 8 Graph the following expressions
a. $x = 3$
b. $y = 2$

To graph these equations, we cannot make an x-y chart since both expressions only contain one variable. Thus, it is important that you know their properties.

SOLUTION

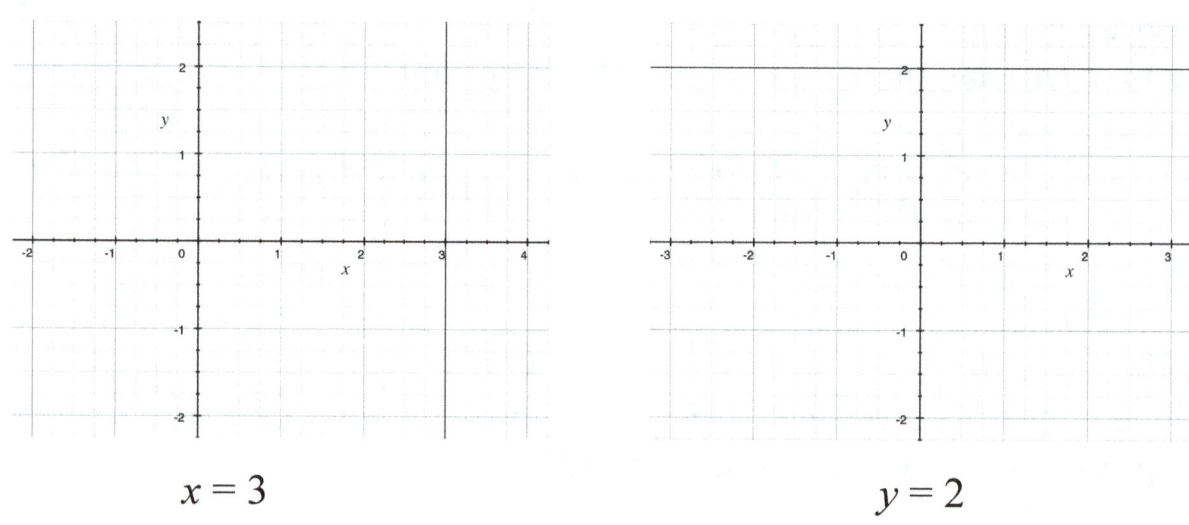

$x = 3$ $\qquad\qquad\qquad\qquad$ $y = 2$

By scrutinizing these two graphs, we found that whenever x is directly equal to a number, it will always be a vertical line perpendicular to the x-axis. Similarly, whenever y equals to a constant, the line will be horizontal and be perpendicular to the y-axis.

Now to calculate the slope of the line in example 8. For $x = 3$, we pick points $(3, 0)$ and $(3, 1)$. Knowing that slope equals the change in y divided by change in x, we obtain: $(1 - 0)/(3 - 3) = 1/0 = \emptyset$. We found that the line's slope is undefined!

However, for $y = 2$, we pick points $(0, 2)$ and $(1, 2)$. Similarly, we obtain: $(2 - 2)/(1 - 0) = 0/1 = 0$. Thus, the slope of a horizontal line is always 0.

Property 1-3   The slope of a vertical line is **undefined**.

Property 1-4   The slope of a horizontal line is **always 0**.

Example 9   Graph the lines and determine their relation between each other:
a. $f(x) = 2x + 1$ and $g(x) = 2x + 2$
b. $f(x) = x + 1$ and $g(x) = -x + 1$

SOLUTION

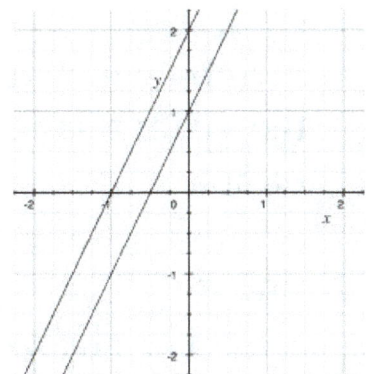 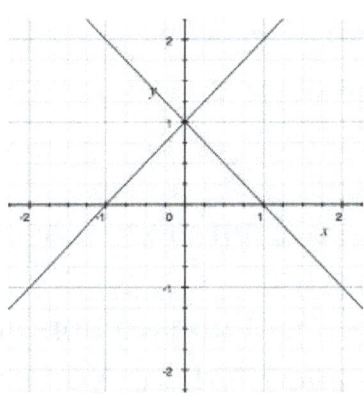

With these graphs, we found if two functions have the same slope, then they are parallel. The opposite, if two functions' slopes' product is –1, then these two lines are perpendicular to each other.

Exercise 5   Graph these functions and determine if they are parallel or perpendicular.
    a. $f(x) = 2x + 1$ and $g(x) = 5x + 1$
    b. $f(x) = -4x + 3$ and $g(x) = 1/4x + 2$
    c. $f(x) = 6x + 5$ and $g(x) = 6x + 2$

1.2 Vocaulary and Excerise

A. Fill in the blank with proper terms

1. Name the following relationship:

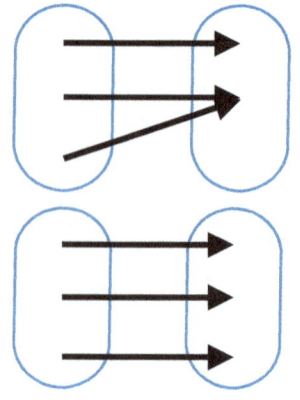

2. A function is the relationship with _____.

3. With two non-empty sets A and B where $A \subset \mathbb{R}$ and $B \subset \mathbb{R}$. Given a method f where $f: A \to B$. Therefore, every element in A has one and one only corresponding element in B. If f satisfied this definition, f is a _____.

4. Slope is a _____.
5. Slope(m) = _____.
6. A _____ is the variable that is the result of the function.
7. An _____ is the input of the function.
8. Y-intercept is the point which the graph of the function _____(touches/intersect) with the y-axis.
9. The slope-intercept form of a function is _____.
10. The point-slope form of the function is _____.
11. Identify the following transformation:

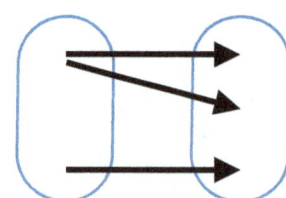

   a. $f(x) = x \pm a$
   b. $f(x) = ax$, where $a \neq 0$

12. A _____ is the method which to test wheter the given relation is a function or not.
13. The slope of a **vertical** line is _____.
14. The slope of a **horizontal** line is _____.

B. Graph the following functions and find the slopes of the line

1. $f(x) = 6x$
2. $f(x) = x$
3. $f(x) = 2x + 1$
4. $g(x) = -2x + 1$
5. $h(x) = (1/2)x$
6. $f(x) = -(½)x$
7. $k(x) = 2x + x + 3$
8. $n(x) = |x|$

9. $f(x) = x/4$

C. Determine what transformation is done in the exercise 2 - 10

D. Is the line parallel or perpendicular? What is the slope of each line?
1. $f(x) = -2x$ and $f(x) = -6x$
2. $f(x) = -2x$ and $f(x) = 2x$
3. $f(x) = -2x$ and $f(x) = -2x + 8$
4. $f(x) = -x$ and $f(x) = 2$

1.3 Systems of Equations

In the previous section, we have discussed the relationships of two lines. Each expression of the lines can be written in the form of an equation with two variables.

Definition 1-31
The basic form of an **equation with two variables** is $ax + by = c$.

We found it is impossible to solve this equation even if substituting a, b, and c with actual numbers. However, if adding another equation, creating a system of equations with two variables.

Definition 1-32
A **system of equations with two variables** has the basic form of: $\begin{cases} ax+by=c \\ dx+ey=f \end{cases}$

A common type of question will be asking if an ordered pair is the solution of the system of equations. Consider the following example:

Example 1 Is the ordered pair (3, 6) a solution to the system of equations $\begin{cases} x+y=9 \\ 2x+3y=24 \end{cases}$?

Key: by substituting the ordered pair into the equations, find if the equation is true.

SOLUTION
Given the ordered pair (3, 6) and the system of equations:
$$x + y = 9:$$
$$3 + 6 \stackrel{?}{=} 9 \qquad\qquad substituting\ numbers.$$

∵ 9 = 9   *equal means the pair satisfies the equation.*
∴ (3, 6) satisfy equation 1
$2x + 3y = 24$:
$2(3) + 3(6)$
$= 6 + 18$
$= 24$
∵ 24 = 24
∴ (3, 6) satisfy equation 2
Therefore, (3, 6) is a solution of the system of equations.

In order to verify that an ordered pair is a solution of the system of equations, it *must* satisfy both equations.

Exercise 1   Is the ordered pair (1, 5) the solution to the system of equations: $\begin{cases} 2x+y=6 \\ 4x-y=-1 \end{cases}$?

___

Solving a system of equations is easy. There are three methods that are applicable to the majority of cases.

Definition 1-33
A **substitution method** is starting from one equation, solving for one variable, then substituting all the variables in the second equation with the first equation.

Consider the following example:

Example 2   Solve the system of equations $\begin{cases} 3x+5y=25 \\ y=2x+1 \end{cases}$ By substitution method.

Key: given the other equation is in the form of a variable equals to an equation. We are able to substitute always in the first equation.

SOLUTION
Given the equation system:

$3x + 5(2x + 1) = 25$ *Substitute y with 2x – 1*
$3x + 10x + 5 = 25$ *Distribute*
$13x = 20$
$x = 20/13$ *Simplify*

$y = 2(20/13) + 1$ *Solve for y*
$y = 40/13 + 1$
$y = 53/13$

∴ the solution is (20/13, 53/13)

However, many systems may not be able to do substitution such as $\begin{cases} 3x-4y=7 \\ 10x+20y=-10 \end{cases}$. Therefore, **elimination method** is advised when solving such systems.

Definition 1-34
Elimination Method is the method when solving a system of equations that first eliminates one variable. Converting the equation into one variable, then solving.

Example 3   Solve the system of equations $\begin{cases} 2x+y=1 \\ 3x-y=4 \end{cases}$.

Key: by adding the two equations, we eliminate y and solve for x. The we evaluate the polynomial (equation) with x.

SOLUTION
with the system of equations: $\begin{cases} 2x+y=1 \\ 3x-y=4 \end{cases}$

$$(2x + y) + (3x - y) = 1 + 4$$
$$2x + y + 3x - y = 5$$
$$5x = 5$$
$$x = 1$$

$$2(1) + y = 1$$
$$2 + y = 1$$
$$y = -1$$

∴ the solution of the system of equations is $(1, -1)$

Look closer at the system of equations. Like what mentioned earlier, all lines can be written into an equation with two variables. Thus, we are able to solve for y in both equations and graph it. The intersection of the two lines is the solution in an ordered pair (coordinate point).

However, this draws situations that are special. Imagine two lines in a coordinate plane. What situations may occur? The most basic situation is that two lines intersect at one point where the solution lies. There may occur that the lines are parallel to each other, causing a null set (no solution). Additionally, lines may be the same and that represents all points on the line that satisfies the equation. For all of the situations, the method of writing results is different.

Property 1-5 When two lines **intersect at one point**: (x, y)

Property 1-6 When lines are **parallel** to each other: ø

Property 1-7 When the lines are the **same**: $\{(x, y) \mid ax + by = c\}$

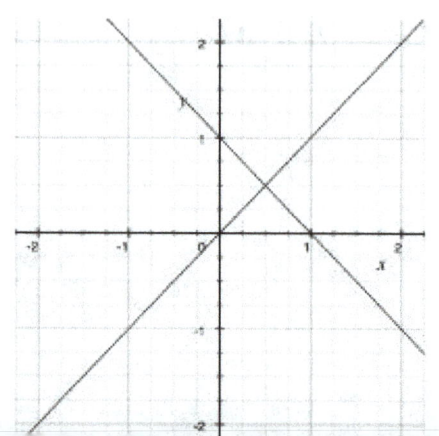 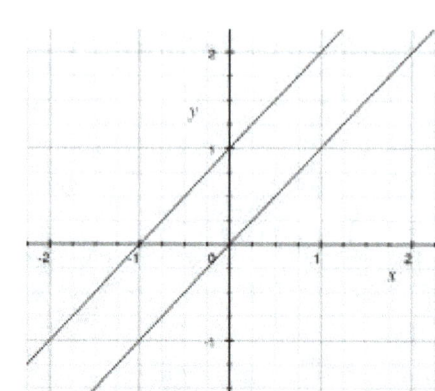 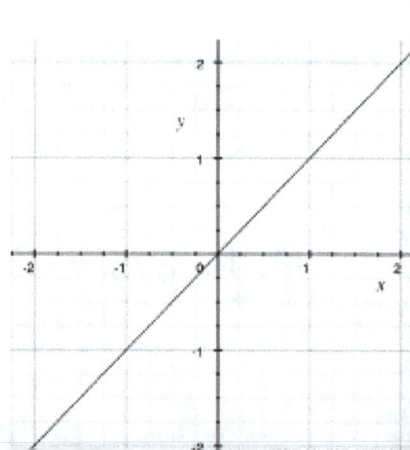

Example 4 Solve the systems of equations below:

a. $\begin{cases} x+y=1 \\ x-y=4 \end{cases}$

b. $\begin{cases} x=y \\ x-y=1 \end{cases}$

c. $\begin{cases} x+3=y \\ y-3=x \end{cases}$

SOLUTION

a. With the system of equations:
$$(x + y) + (x - y) = 5$$
$$x + y + x - y = 5$$
$$2x = 5$$
$$x = 5/2$$

$$5/2 - y = 4$$
$$-y = 4 - 5/2$$
$$y = 5/2 + 4$$
$$y = 13/2$$

∴ the solution of the system of equations are (5/2, 13/2)

b. With the system of equations:
$$y - y = 1$$
$$0 \neq 1$$

∴ the solution of the system of equations is a null set.

c. With the system of equations:
$$(x + 3) + (y - 3) = x + y$$
$$x + 3 + y - 3 = x + y$$
$$x + y = x + y$$

∴ the solution of the system of equations are all points on the line:
$$\{(x, y) \mid x + 3 = y\}.$$

Now, further examine the line equation of all three systems. We discovered that when two lines create a null set, they are parallel, which means their slopes are the same. In this scenario, the result will be a false statement such as $0 = 1$. When $\{(x, y) \mid ax + by = c\}$ is found, it will always create a true statement such as $x + y = x + y$.

Exercise 2 Graph the following systems of equations and solve them. Specify all special cases.

a. $\begin{cases} 3x+y=1 \\ x-2y=5 \end{cases}$

b. $\begin{cases} x+y=1 \\ y=x \end{cases}$

A traditional application of the system of equations with two variables is "caging problems". Consider the following example:

Example 5 There are rabbits and chickens in a cage. From the top, there are 35 heads and 94 feet. How many chickens and rabbits are there?

Key: To solve this question, we first by assuming that all chickens have 2 legs (feet), and rabbits have 4 legs (feet). Since all animals only have one head, there are in total 35 animals in the cage. Based on this information, we were able to write a system of equations.

SOLUTION
Let the number of chickens be x and the number of rabbits be y.

$$\begin{cases} x+y=35 \\ 2x+4y=94 \end{cases}$$

$$x = 35 - y$$
$$2(35 - y) + 4y = 94$$
$$70 - 2y + 4y = 94$$
$$2y = 24$$
$$y = 12$$

$$x = 35 - 12$$
$$x = 23$$

Thus, there are 23 chickens and 12 rabbits in the cage.

Exercise 3 There are several balls in a box. The bigger ones weighted 20g and the smaller ones weighted 10g. Without the box, all balls weight 250g. How many big and small balls are there?

_____

1.3 Vocabulary and Exercise

A. Fill in the blank with proper terms

1. The basic form of an equation with two variables is _____.

2. The basic form of system of equations is _____.

3. Substitution method is to _____.

4. Elimination method is to _____.

5. When two lines intersect at a point, you write the solution as _____.

6. If the lines _____, you write as a null solution set.

7. When two lines are the same, you write as _____.

B. Determine if the ordered pair is the solution of the systems of equations

1. $(1, 5)$ $\begin{cases} x+y=35 \\ xy=5 \end{cases}$

2. $(2, 8)$ $\begin{cases} -xy=15 \\ x+y=10 \end{cases}$

3. $(8, 9)$ $\begin{cases} \frac{x}{y}=-1 \\ x=8 \end{cases}$

4. $(10, 12)$ $\begin{cases} 2x+4y=68 \\ -xy=120 \end{cases}$

5. $(-1, -5)$ $\begin{cases} x+y=-6 \\ xy=5 \end{cases}$

6. $(3, 7)$ $\begin{cases} 3(x+y) - 3=15 \\ 6(xy)+4=22 \end{cases}$

7. $(6, 5)$ $\begin{cases} x+y+1=11 \\ x/xy=5 \end{cases}$

8. $(0, 0)$ $\begin{cases} x+y-5=-6 \\ xy=-1 \end{cases}$

9. $\emptyset$ $\begin{cases} x+y=6 \\ xy=5 \end{cases}$

10. $(4, 4)$ $\begin{cases} 2+y=6 \\ x+y-5(3xy)=5 \end{cases}$

C. Solve the system of equation

1. $\begin{cases} x+y=6 \\ 3x-6(x+2y)=5 \end{cases}$

2. $\begin{cases} x-y=-7 \\ 5x-4y=5 \end{cases}$

3. $\begin{cases} x+y=8 \\ x-y=-3 \end{cases}$

4. $\begin{cases} x+y=6 \\ x(y-4)=5 \end{cases}$

5. $\begin{cases} 2y=6 \\ 3y-8x+50=100 \end{cases}$

D. Solve

1. There are motorbikes and cars in the parking lot. There are in total of 130 wheels and 70 vehicles in total. How many motorbikes and cars are there?

Chapter 1 Summary and Review

A. Vocabulary
1. Monomials
2. Polynomials
3. Equation
4. Inequality
5. Terms
6. Degree
7. Coefficients
8. Like terms
9. One-to-one relationship
10. One-to-many relationship
11. Many-to-one relationship
12. Function
13. Ordered pair
14. Independent variable
15. Dependent variable
16. Slope
17. y-intercept
18. Slope-intercept form
19. Point-slope-form
20. General form
21. Parent function
22. Transformation
23. Vertical Line Test
24. Equation with two variables
25. System of equation with two variables
26. Substitution method
27. Elimination method

B. Key Concepts
1. Monomial is the product of numbers and variables; letters and numbers alone are also monomials.
2. A polynomial is a series of monomials joined by addition and subtraction.
3. An equation is a monomial or polynomial that equals to another.
4. Inequality is a certain relationship where two monomials are *not* joined by an equal symbol.
5. A polynomial is two or more monomials that are joined by addition or subtraction.
6. Each monomial in a polynomial is called a term.
7. The greatest power of any term is the degree of the polynomial.
8. Any term that only consists of a number is called a constant.
9. All numbers multiplied by a variable are called coefficients.

10. Terms in polynomials have the same variable and power are called like terms.

11. An equation where one input has one and one only output is called a function.

12. The slope is the rate of change in a function. Represented by letter.

13. Slope will always be equal to the change in y-value divided by the change in x-value.

14. A dependent variable is the variable that is the result of the operation(output).

15. An independent variable is the input of the function.

16. The y-intercept is the point where the graph of the function intersects with the y-axis.

17. A Slope Intercept Form is often expressed as $f(x) = mx + b$. In this format, m is the slope and b are the y-intercept.

18. A Point Slope Form is the form that uses a point on the line and the slope to define the line.

19. A point slope form is written as: $y - y_1 = m(x - x_1)$

20. A general form is another variant of a form of a linear equation.

21. The general form is $a + by = c$

22. All functions will have a parent function which is simply the most basic form of a function. A parent function's expression is a monomial.

23. $f(x) = x + a$ shift the function up by a units

24. $f(x) = x - a$ shift the function down by a units

25. $f(x) = ax; a \neq 0$ when a > 0, this will increase the slope of the function where the line will get steeper. When $a < 0$, this will flip the graph. This flipped graph is symmetrical to the original graph by y-axis.

26. A vertical line test is the method which tests whether the given relation is a function or not.

27. The slope of a vertical line is undefined.

28. The slope of a horizontal line is always 0.

29. If two functions have the same slope, then they are parallel. The opposite, if two

functions' slopes' product is – 1, then these two lines are perpendicular to each other.

30. The basic form of an equation with two variables is $ax + by = c$

31. A system of equations with two variables has the basic form of:
$\begin{cases} ax+by=c \\ dx+ey=f \end{cases} (a, b, c, d, e, f \in R)$

32. A substitution method is starting from one equation, solving for one variable, then substituting all the variables in the second equation with the first equation.

33. Elimination Method is the method when solving a system of equations that first eliminates one variable. Converting the equation into one variable, then solving.

34. When two lines intersect at one point: (x, y)

35. When lines are parallel to each other: ø

36. When the lines are the same: $\{(x, y) \mid ax + by = c\}$

Chapter 1 Excercise Answer Key

1.1 Exercise
1. a~d
   a. monomial
   b. Polynomial
   c. Inequality
   d. Equation
2. 3rd degree; constant: 5; coefficient: 1, 1, –1
3. a~b
   a. $81 + 162 + 21 + 24 = 288$
   b. $25 + 85 + 3 = 113$
4. $x = 3$
5. $10x + 3000 = 5000$
   $x = 200$
   $100(200) + 3000 = 23000$

1.2 Exercise
1. $m = 3$
2.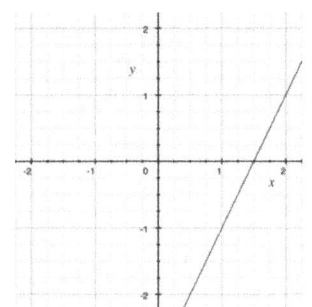

3. $f(x) = (½)x + 1$
4. reflected by the y-axis, more sleeper($m = 5$), up 12 units
5. No
6. a~c
   a. None
   b. Perpendicular
   c. Parallel

1.3 Exercise
1. No

2.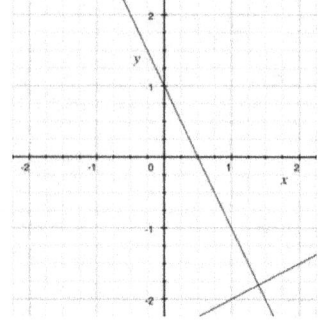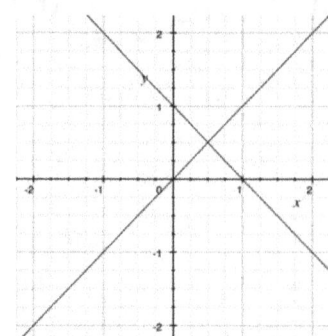

3. 10 Large and 5 small

Chapter 2 Polynomials and Quadratics

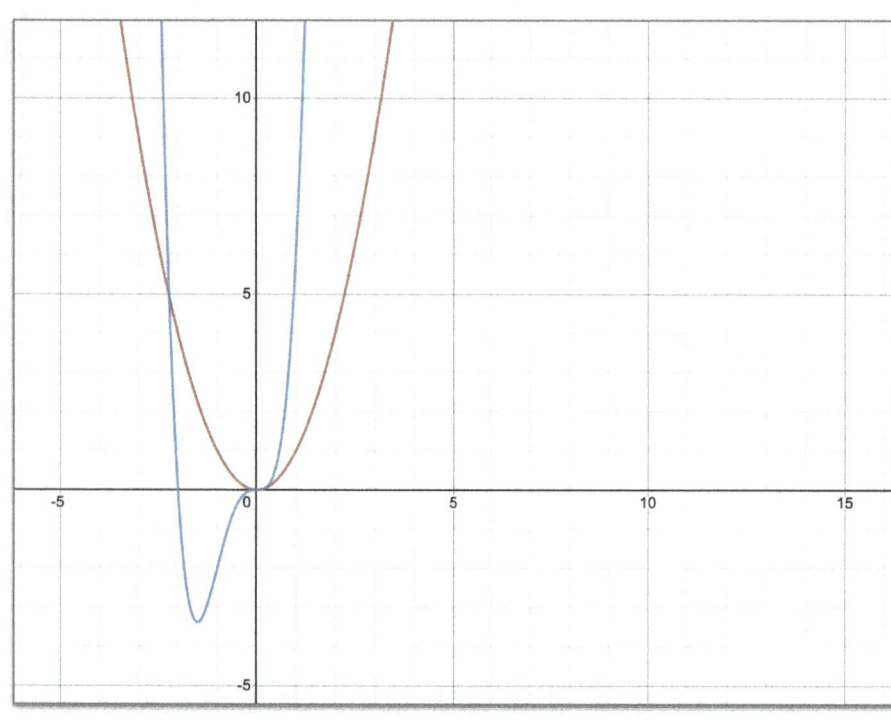

The graph of a quadratics and a polynomial function. $f(x) = x^2$ and $f(x) = 2x^4 + 4x^3$
In this chapter, we will be discussing the properties and basic operations with polynomials. Additionally with the basic knowledges from quadratics.

2.1 Operations with Polynomials and Factoring

In the previous chapter, we discussed the definition of a polynomial. Like other elements in math, we can perform most mathematical operations with polynomials. These operations tie in with all numeral operations which you should already have mastered. Consider the example:

Example 1 Evaluate the following
  a. $3(2 + 1)$
  b. $(5 - 3) - (4 - 2)$

SOLUTION
  a. $3(2) + 3(1) = 6 + 3 = 9$
  b. $5 - 3 - 4 + 2 = 2 - 4 + 2 = -2 + 2 = 0$

If you do not recognize any of the patterns above, or are not able to perform the corresponding operation, **remember the following rules**:

Property 2-1		$a + b + c = (a + b) + c$
Property 2-2		$a \cdot b \cdot c = c(ab)$
Property 2-3		$a(b + c) = ab + ac$

The rules above are applicable for all operations, including polynomials. With all the above, we are able to perform any polynomial operations. Thus, consider the following: $x(2x + 3)$. If we are asked to expand this, what rule should we use? The third rule $a(b + c) = ab + ac$ will be used. Imagine x as an, and $2x + 3$ as $b + c$. Then we can expand this into: $2x^2 + 3x$. However, consider expanding the monomial: $(2x + 1)(3x + 2)$. When expanding this polynomial, see one of the sub-terms, and there you obtain:

$$\text{Original} = 3x(2x + 1) + 2(2x + 1) \quad \textit{expand by third rule}$$
$$= 6x^2 + 3x + 4x + 2$$
$$= 6x^2 + 7x + 2 \quad \textit{Simplify}$$

Based on the proof above, we can identify a pattern when having two polynomials multiplying each other:

1) $(n + n_1 + n_2 + n_3 + \ldots + n_n)(n + n_1 + n_2 + n_3 + \ldots + n_n) = n(n + n_1 + n_2 + n_3 + \ldots + n_n) + n_1(n + n_1 + n_2 + n_3 + \ldots + n_n) + \ldots + n_n(n + n_1 + n_2 + n_3 + \ldots + n_n)$
2) Any two polynomials multiply together equal to **every term** in the first polynomial multiply by the second polynomial.

In elementary school, we learned that division is equivalent to multiplying the reciprocal of the number. Such that: $a \div b = a(1/b)$. Similarly, when operating with polynomials and monomials, we can multiply one polynomial by the other's reciprocal. For example:

$$3x \div (6x - 3)$$
$$= 3x[1/(6x - 3)] \quad \textit{multiply the first by the reciprocal}$$
$$= 3x/(6x - 3)$$
$$= 3x/3(x - 1)$$
$$= x/(x - 1) \quad \textit{Simplify}$$

Performing basic addition and subtraction of polynomials should be simple at this moment for you. All key information that you should acknowledge at this point are:

1) $a + (b - c) = a + b - c$
2) $a + (b + c) = a + b + c$
3) $a - (b + c) = a - b - c$
4) $a - (b - c) = a - b + c$

The concept remains when operating with polynomials. Notice that when changing the signs, all signs must be changed!

YES: $\quad a - (b - c + e - k) = a - b + c - e + k$

NO: ~~a − (b − c + e − k) = a − b + c + e + k~~

Middle school math had taught you how to perform a basic exponential operation. Now you should understand and be able to identify each component in an exponential expression:

$$a^n = a \cdot a \cdot a \cdot ... \cdot a$$

In the expression shown above, *a* is the base and *n* is the exponent. Additionally, you should always know that an exponential expression is the simple form of *n* a(s) multiplying. With a polynomial, the definition remained. Consider the following example:

Example 2 Expand the following monomials
 a. $(x + 3)^2$
 b. $(x - 3)^2$

Key: write the exponential expression in its full form and based on what we learned, expand the polynomial.

SOLUTION
 a. For $(x + 3)^2$, we rewrite into $(x + 3)(x + 3)$ from here, we expand the polynomial:
$$x(x + 3) + 3(x + 3)$$
$$= x^2 + 3x + 3x + 9$$
$$= x^2 + 6x + 9$$
 b. For $(x - 3)^2$, write into $(x - 3)(x - 3)$, then expand:
$$x(x - 3) - 3(x - 3)$$
$$= x^2 - 3x - 3x + 9$$
$$= x^2 - 6x + 9$$

From the process above, we discovered a pattern that was repeating whenever we expanded a squared two-term polynomial. This pattern is called the **Perfect Square**. Practice more about expanding a perfect square since it is critical in the second section.

Property 2-4 $\qquad (x \pm y)^2 = x^2 \pm 2xy + y^2$

Exercise 1 Expand: $(a - b)^2$

___

We have learned in middle school that exponentials have properties that handle operations among two or more exponentials. Recall the knowledge:

Property 2-4 $\qquad (a^b)^c = a^{bc}$

Property 2-5 $\qquad (a^b)(a^d) = a^{b+d}$

Property 2-6 $\qquad (a^2 - b^2) = (a + b)(a - b)$

Based on this, we can expand these rules to apply to polynomials. By having two squared polynomials multiply by each other. Instead of having two large polynomials subtracting each other, having two of them first add and subtract, then multiply.

Exercise 2 Expand: $(x^2 + y^2)(x^2 - y^2)$

___

With the knowledge of expanding a polynomial, factoring is the process of performing the reverse operation from expansion. Be sure to recognize the key patterns that were mentioned in 2.1. Such an example would be: $x^2 + 2xy + y^2$. When factoring, *caution on signs*.

Factoring in polynomials that do not fit any of the special cases above are a little tricky. In fact, the methods that this book uses are difficult to learn.

However, as soon as you have mastered this method. It will be simple to factor in any polynomials.

This method that is suggested is called a **cross-multiplication method**. For example, when factoring the polynomial: $x^2 - 6x - 8$. From the first observation, this does not fit any special cases above. In fact, by using cross-multiplication, we can simplify the process:

1. Set the basic structure of cross multiplication: $\begin{smallmatrix}ax & c\\ bx & d\end{smallmatrix}$. This form when you multiply vertically, it will get the first term. By multiplying c and d with the sign, you will get the last term of the polynomial.
2. Find the correct value for c and d when you multiply diagonally and add the product, the sum will be the middle term.
3. When writing the result, write horizontally. If no signs are specified, use addition. The two sums multiply by each other which will obtain the factored polynomial.

Example 3   Factor by cross multiplication: $x^2 - 6x - 7$

Key: follow the steps of the process of cross-multiplications. Structure the corresponding sets.

SOLUTION
with the given polynomial $x^2 - 6x - 7$:

$$\begin{matrix} x & -7 \\ x & 1 \end{matrix}$$

Both a and b are 1 because there is no coefficient in front of the squared term. Multiplying and adding vertically, we obtain: x^2 and -8, which are the first and last terms in the polynomial. This satisfies the requirement of cross-multiplication. Thus, we obtain the factored polynomial: $(x - 7)(x + 1)$. To verify, we perform:

$$x(x + 1) - 7(x + 1)$$
$$= x^2 + x - 7x - 7$$
$$= x^2 - 6x - 7$$

From the example above, we have the first practice of factoring with cross-multiplication. In fact, this method is widely appreciated not only for second-degree polynomials but also for multiple degrees. Such that $x^3 - 6x - 7$. Similarly, we set up the cross-multiplication structure as: $\begin{matrix} ax & c \\ bx & d \end{matrix}$. You can infer that whenever the largest degree's coefficient term does not equal to 1, a and b will not all be one. *Notice that this method **does not** apply to polynomials that exceed more than 3 terms.* For example, $x^3 + 3x^2 - 6x - 7$, the cross-multiplication will not be able to apply.

When encountered in such cases, we use **the factor by grouping** method. Recall the basic rule of monomials operation where $a + b + c + d = (a + b) + (c + d)$. When factoring in such a rule, the following formulas should be remembered:

1) $ab + ac = a(b + c)$
2) $a^2 - b^2 = (a + b)(a - b)$
3) $a^3 - b^3 = (a - b)(a^2 + ab + b^2)$
4) $a^3 + b^3 = (a + b)(a^2 - ab + b^2)$

Example 4 Factor by grouping: $x^2 + 2x - 3x - 6 - 2x^2 + 3x + 9$

Key: first combine like terms. Then use the appropriate formula that fits this case.

SOLUTION
with the original:

$$x^2 + 2x - 3x - 6 - 2x^2 + 3x + 9$$
$$= (x^2 - 2x^2) + (2x - 3x + 3x) - (6 - 9)$$
$$= -x^2 + 2x + 3$$
$$= -(x^2 - 2x - 3)$$
$$= -(x - 3)(x + 1)$$

In the example above, we factored the negative out of all terms to continent the process. Factoring is critical to the following lessons in the future, when solving a quadratic, one of the methods is to factor, then solve.

Exercise 3 Factor: $x^2 + 4x - 7x - 28$

2.1 Vocaublary and Exercise

A. Fill in the blank with proper terms

1. $a^n =$ _____.
2. The perfect square property is _____.
3. The method when factoring polynomials is called _____.
4. If the polynomial has more than _____ terms, we use factor by grouping

B. Expand the following
1. $a(b + c)$
2. $(a + b)(c + d)$
3. $a(a + c)(d - f)$
4. $(a + b)^2$
5. $(a + b)^4$
6. $(a + b + c)^2$
7. $(a + 3)^2(4 + x)$
8. $(a + b)(a - b)$
9. $(x^2 - y^2)(x^2 + y^2)$
10. $(a + b + c - d)^3$

C. Given two polynomials P and O, where $P = x^2 + y$ and $O = x^4 - y$, perform the operation, the question is asked. Factor in the result.

1. $P + O$
2. $P - O$
3. $2(P)$
4. $2(PO)$
5. $-P(O)$
6. P^2
7. $(OP)^2$
8. O^2
9. $-(O)^2$
10. $(PO)^3$

2.2 Quadratic Equation

In the previous lessons, we learned how to factor a polynomial and operations. In this chapter, we will be learning how to graph a polynomial. One specific type of polynomial: quadratics.

Definition 2-1

A **quadratic equation** is an equation with a *squared* term.

The most basic form of the quadratic equation is $ax^2 + by + c = 0$. Such that $x^2 + 2y + 3 = 0$. In the next lesson, we will learn how to solve for x in the next lesson. However, for now, we will only be discussing the polynomial side of the quadratic. Examine the input/output chart below, what is significant about a quadratic polynomial?

Regarding the table, we discover that it has a many-to-one relationship, thus making the relationship a function. If we were to plot those points into a coordinate plane, we would obtain the following:

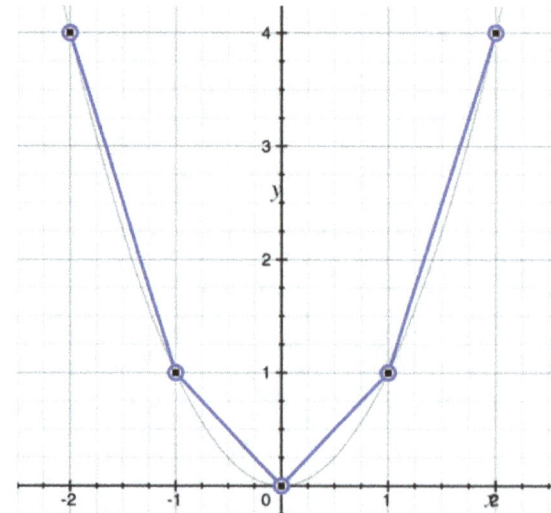

Then, if we connect those points, the lines will connect into a "U" shape. If plotting more lines, you will discover that the line which connects the points will be remarkably close to the fainted line in the graph above. This shape is called a **parabola**.

Definition 2-2
A **parabola** is a U-shaped, y-axis symmetrical figure.

Thus, the definition of parabola suggested that it is y-axis symmetrical where two inputs may have one output. Where in the equation $y = x^2$, both -1 and 1 will be resulting 1. However, this does not disqualify the equation as being a function, since a function cannot have one input resulting in two different outputs.

Example 1 Graph the quadratic equation: $y = x^2 + 3$

To graph a quadratic, we must use the **five-points rule** where we must figure out at least five points before graphing the equation. To find the five points, we will use the x-y chart to find the points, then graph it like a linear equation.

SOLUTION
with the quadratics: $y = x^2 + 3$:

With such points, plot them in the coordinate system, then connect points will smooth line:

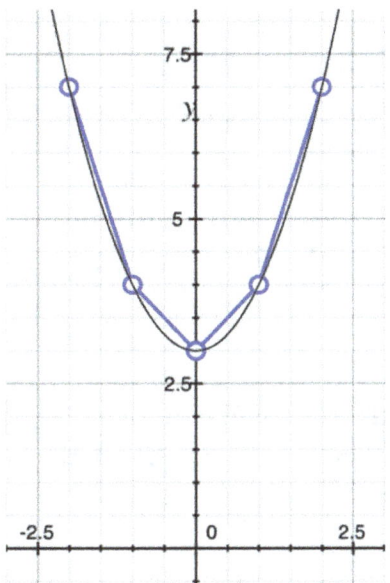

Then you have obtained the graph of $y = x^2 + 3$.

Consider remembering the following points set regarding a quadratic graph: (–2, 4), (–1 , 1), (0 ,0),
(1, 1), and (2, 4). Without any transformation, all quadratic equations will be passing through these points.

As mentioned, all graphs will have a parent graph towards it. For a quadratic, the parent function will be $y = x^2$. This function is which the graph passes through points (–2, 4), (–1 , 1), (0 ,0),
(1, 1), and (2, 4).

Exercise 1   Graph $y = 2x^2$

When having a graph, you should always identify the **domain** and **range** of the function.

Definition 2-3
A **domain** is the range of values for *x*.

Definition 2-4
A **range** is the range of values for *y*.

Definition 2-5
Write domain and range in the interval notation such that: (min, max).

All domain and range in interval notation will represent the max and min values for the function. Brackets indicate that the value is included while parentheses mean the value is not included.

If the value happens to be infinity, use the infinity symbol. However, notice that both negative and positive infinity *cannot be included*. The determination of bracket usage is to see if the value will undefine the function such that: *f(x)*

= 1/x, to write the domain, first think what value of x will make f(x) undefined, since x is on the denominator, if x equals to 0, the function is undefined. Then the domain of f(x) is *everything but zero*. To translate a sentence into symbols, we first determine the minimum value of the function. In this case it is a negative infinity. In fact, the maximum is also a positive infinity.

However, notation (−∞, ∞) will not apply since 0 is included in the domain. Therefore, we must write domain notation separately. By using the union symbol (∪) to join the two parts together, we obtain:
(−∞, 0) ∪ (0, +∞). This applied to the function correctly since the zero is not included within the range of the definition.

Example 2   Determine the range of $f(x) = x^2 + 4$.

Key: find the maximum and minimum value of the function, then write the domain.

SOLUTION
with the quadratic function: $f(x) = x^2 + 4$:

>Max value: infinity
>Min value: 4
>Therefore, the range of $f(x) = x^2 + 4$ is [4, +∞).

In the example above, the point (0, 4) is where the minimum value occurred. Since the parabola passed through the point, it is included.

The point where the maximum or minimum value occurred is called the **vertex**. In a parabola to find the vertex, we use the formula

$$(\frac{-b}{2a}, f(\frac{-b}{2a}))$$

In this formula, we introduced the standard form of a quadratic function: $f(x) = ax^2 + bx + c$. By substituting the values in the formula with actual values, we can obtain the vertex.

Previously, we briefly discussed the parent function of a quadratic which is $f(x) = x^2$. Like linear functions, the transformations of these quadratics are:

Property 2-7   If the coefficient of the squared term is positive, then it has a minimum

Property 2-8   If the coefficient of the squared term is negative, then it has a maximum

Property 2-9   Values that are added or subtracted from x in squared terms will move horizontally.

Property 2-10   Constants added or subtracted will move vertically.

It is critical to know how to identify the vertex of a parabola without having to graph. When finding the range of a function, the minimum or maximum value is where the vertex is located. Consider the following example:

Example 3   Determine the vertex of $f(x) = x^2 - 4$.

Key: use the formula for the finding the vertex of a parabola.

SOLUTION
with the given function: $f(x) = x^2 - 4$:

$$V = (0/2, f(0/2))$$
$$= (0, f(0))$$
$$= 0^2 - 4$$
$$= -4$$

Therefore, the vertex of the quadratic function $f(x) = x^2 - 4$ is at $(0, -4)$

Previously, we have mentioned that the vertex is where the maximum or minimum value lies. Recall how range defines a parabola. Since a parabola's range of y-values will always be [y-coordinate of vertex, ∞). When finding the range of a parabola, therefore, for any given function (mapping) f, with a positive coefficient on the squared term, its range's maximum value will always be positive infinity.

With domain, imagine the graph of $f(x) = x^2$. The ends of both sides are all arrows which indicate that it is continually expanding. Regardless of any transition, both ends will never stop expanding. Hence, the domain of *any* quadratic function is always $(-\infty, \infty)$.

Exercise 2   Determine the domain and range for: $f(x) = 2x^2 + 3x - 4$.

Since now, all equations and functions that we discussed are all in the form of $y = ax^b + cx + d$. However, when we graph a function where x and y variables have been switched and compare the new graph of the function to the original, what are the differences? To discuss that, we start by graphing a quadratic equation: $y = x^2$ in a coordinate plane (shown on right)

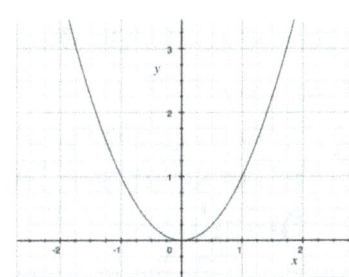

Then, by switching the x and y variables, we obtain the inversed equation:

$x = y^2$. However, a normal calculator will not have the ability of directly graphing this equation. Therefore, we must solve for y. To obtain y, we must take the square root of both sides thus transform this equation into: $\sqrt{x} = \sqrt{y^2}$ further simplify the equation into: $\sqrt{x} = y$. Hence the equation has transformed into: $y = \sqrt{x}$. Then graph the equation in the same coordinate plane(shown on right).

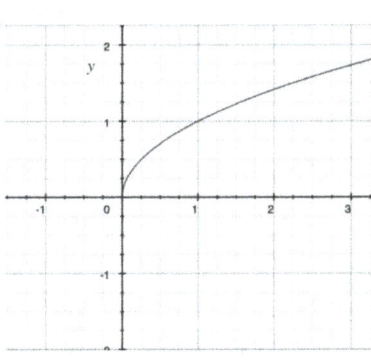

We notice that the graph is missing the bottom section of the parabola. Since $\forall x < 0: \sqrt{x}$ is undefined, if $x = -1$, $\sqrt{-1} = i$, the imaginary number, and it is impossible to graph an irrational number(imaginary number) on a coordinate plane. Therefore, the bottom section of the parabola will not appear.

An imaginary number, as what its name suggested, can only be imagined. We use i to represent this number. Imaginary numbers can perform certain operations including exponentials. Since the value of i is $\sqrt{-1}$. We can obtain the value of the square of i as -1. With mathematical symbols i can be represented as $i \notin \mathbb{R}$.

However, consider the graph on the right, if there is an expression for the graph, will the expression be a function? The self-evident that it is not a function. In fact, the expression of the graph above is $x = y^2$. This equation appeared in the process of obtaining the inverse of $y = x^2$. The equation $x = y^2$ has the property which $\exists x \notin \mathbb{R}$ and $\exists y \notin \mathbb{R} : x = y^2$ defined. Hence when graphing the equation, both sides of the parabola is graphed.

The process that obtaining the inverse of a function is called taking the **inverse of the function**. The product of the operation is called **inverse function**.

Definition 2-6
The inverse of the function is first switching the function's dependent and independent variable then solve for dependent variable.

Definition 2-7
The notation of an inverse function is $f^{-1}(x) = \ldots$

In elementary school, we learned that by taking the -1 power to a number, we obtain the reciprocal of the number, such as: $3^{-1} = 1/3$. Similarly, by "taking the -1 power" of a function, we have the inverse function. Additionally, we learned that the product of reciprocal and the original is always 1. Therefore, this implies that by performing certain operation with two function that are inversed, we can obtain 1 as the result. The process that we take is **composition**.

Definition 2-8

With A_1 and A_2 are two non-empty sets. $\exists f_1$ and $\exists f_2$ as two methods which satisfy: $f_1: A_1 \to E_1$ and $f_2: E_1 \to B$. When $E_1 \subset B$, from f_1 and f_2, we can obtain a method where it will map an element $x \in A_1$ into $f_1(f_2(x)) \in B$. Then, we call the mapping from A_1 to B a **composite mapping**.

Definition 2-9

With the definition of compositional mapping, we obtain the definition of **composite function** as a special case where function notation used for composition mapping: with two functions: $f(x)$ and $g(x)$. With the domain of $f(x)$ as D_f and $g(x)$ as D_g, the range of $g(x)$ as R_g where $R_g \subset D_f$, then we call the function: $f(g(x))$ as a composite function.

To verify that if a function is the other's inverse function, we take the composite of $f(x)$ and $f^{-1}(x)$. When the result be x, then both functions are inverse of each other.

Example 4 Determine if $f(x) = x^4 + 1$ and $g(x) = \sqrt[4]{x-1}$ are inverse functions.

Key: Take the composite of the functions and see if the result is x.
Solution: with $f(x) = x^4 + 1$ and $g(x) = \sqrt[4]{x-1}$:
$$f(g(x)) = (\sqrt[4]{x-1})^4 + 1$$

$$= (x-1) + 1$$
$$= x$$

Since the result of $f(g(x))$ is x, then $f(x)$ and $g(x)$ are inverse functions.

2.2 Vocabulary and Exercise

A. Fill in the blank with proper terms

1. A _____ is a equation with a squared term.
2. A parabola is a _____.
3. The domain of a function is _____.
4. The _____ of a function is the range of the possible y values.
5. We often write domain and range of a functions as _____.
6. If the coefficient of the squared term is positive, then the function has a _____.
7. If the coefficient of the squared terms is _____, then the function has a maximum.
8. A constant added(subtracted) to the end of the function will move _____.
9. A constant added(subtracted) to the squared term will _____.
10. An inverse function is when we _____.
11. The notation of an inverse function is _____.

B. Determine if the expressions are quadratics and explain why

1. $y = 2k$
2. $y = x^2$
3. $y = 2x^2/x^2$
4. $y > 3x^2$

C. First to find the vertex of the following quadratics, then graph and determine the domain and range.

1. $f(x) = x^2$
2. $f(x) = x^2 - 2$
3. $f(x) = (x-3)^2$
4. $f(x) = (x+3)(x-3)$
5. $f(x) = 1/2(x+3)^2$
6. $f(x) = 2x^3/x$

D. Find the inverse of the following functions.

1. $f(x) = x^4$
2. $f(x) = 1/2x^2$
3. $g(x) = (1/x) - 2$
4. $g(x) = {}^{32}\sqrt{x} + 1$
5. $h(x) = x^{114} - 3$
6. $h(x) = x^2 + 32x + 256$
7. $k(x) = 2x^{514} + 6$
8. $k(x) = \sqrt{2x} + 4$
9. $p(x) = (2x)^2$
10. $p(x) = 1/(x-1)$

2.3 Solving Quadratics

All equations will have a solution no matter what. For any linear equations, the solution will be the x-intercept of the line. Similarly, for a non-linear equation such as a parabola, the solution will also be the x-intercept. In fact, imagine the equation $y = x^2 - 1$. The graph intersects the x-axis at –1 and 1. Since at both points' y-value is 0, we set the y variable to 0 and substitute x by either –1 or 1. We conclude that the equation is valid. The process is called to solve a quadratics.

Majorly there are three methods to solve a quadratics.

1) Factoring
2) Graphing
3) Quadratic equation

The first method, factoring, is by set the equation into zero, then factor the other side to solve the equation. Consider the following the example:

Example 1   Solve: $y = x^2 - 1$.

SOLUTION
Since the solutions' y value is 0, we set the equation into 0:
$$0 = x^2 - 1$$
$$(x + 1)(x - 1) = 0$$
$$x = -1$$
$$x = 1$$

When solving for x is first by examine the factored equation $(x + 1)(x - 1) = 0$. To result this equation to 0, either $(x + 1)$ must equals to 0 or $(x - 1) = 0$. With these two equations, we solve for x, which will find the solution.

Exercise 1   Solve: $x^2 + 2x = -1$

Another method is to graph the equation into a coordinate plane. As what mentioned, the equation's solution will be the x-intercept of the equation. However, notice that equation
$y = x^2 + 1$ will not have a x-intercept since it is shifted up by 1 unit. To further proof the conclusion, we set the equation equals zero then solve.

$$x^2 + 1 = 0$$
$$x^2 = -1$$
$$x = i$$

Since the solution is not a real number, we say that the equation does not have a real root. A **root** in this context meaning the solution of a quadratic equation.

Example 2   Graph then solve for the equation: $(x - 3)^2 = 2$.

SOLUTION
To graph the equation, we first determine the parent function which is $y = x^2$. By transformation, we know that this equation being shifted down two units and right 3. Since there is no coefficient for x, then we graph the equation.

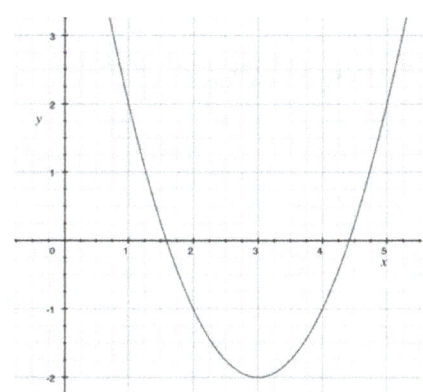

By observing the equation, we found the x intercept of the parabola is 1.586 and 4.414. therefore, the solution of the equation is 1.586 or 4.414.

If there happens to not have any intersection with the x-axis, such as equation $y = x^2 + 1$, then we conclude that the equation does not have any real root.

Exercise 2   Graph then solve: $(x - 5)^2 + 3 = 0$

___

To determine if the solution of a quadratic is a real number, we use the **delta discriminant**. First, recall the basic form of a quadratic equation: $y = ax^2 + bx + c$. The three non-variable values: $a, b,$ and c are the key values of the discriminant.

Definition 2-10
$$b^2 - 4ac = \Delta.$$

The usage of the discriminant is defined by the following definition:

Definition 2-11
$$b^2 - 4ac < 0 \Leftrightarrow \frac{-b \pm \sqrt{b^2 - 4ac}}{2a} \notin \mathbb{R}.$$

By using the delta discriminant, we can construct a formula to solve any quadratic equations. We first by setting up the basic form of a quadratic equation: $ax^2 + bx + c = 0$. Our goal in this process is to isolate x. Therefore, we can have the process below:

Given that $ax^2 + bx + c = 0$:

$$x^2 + \frac{bx}{a} + \frac{c}{a} = 0$$

$$x^2 + \frac{bx}{a} = -\frac{c}{a}$$

$$x^2 + \frac{bx}{a} = -\frac{c}{a}$$

$$x^2 + \frac{bx}{a} + \frac{b^2}{4a^2} = -\frac{c}{a} + \frac{b^2}{4a^2}$$

$$(x + \frac{b}{2a})^2 = \frac{(-4ac + b^2)}{4a^2}$$

$$x + \frac{b}{2a} = \frac{\pm\sqrt{-4ac + b^2}}{2a}$$

$$x = \frac{-b \pm \sqrt{b^2 - 4ac}}{2a}$$

The formula that we just concluded is called the **quadratic formula**. This is applicable for all quadratic equations. Now look closer to the parts under the square root. It is the delta discriminant. Hence, this explained that the application definition: $b^2 - 4ac < 0 \Leftrightarrow x \notin \mathbb{R}$.

Usually, when we take the square root of a number, we are referencing to a normal square root. The definition of that will be: $\forall x \subset \mathbb{N}: \pm\sqrt{x}$ is always defined. The operation that explained in this definition is called a square root. The ± sign suggested that there will be two results, positive and negative. In fact, the **principle square root**, shown as: $\forall x \subset \mathbb{N}: \sqrt{x}$ is always defined. The principle square root will only have the positive result.

Example 3   Solve the equation: $x^2 + 3x - 5 = 0$

Key: use the quadratics formula to solve the equation

SOLUTION

$$x = \frac{-b \pm \sqrt{b^2 - 4ac}}{2a}$$
$$x = -3 \pm (3^3 - 4(1)(-5))^{1/2}/2$$
$$= -3 \pm (9 + 20)^{1/2}/2$$
$$= (-3 \pm 29^{1/2})/2$$
$$\approx (-3 \pm 5.39)/2$$
Solution 1: $2.39/2 = 1.195$
Solution 2: $-8.39/2 = 4.195$

The solution of the equation is $x \approx 1.195$ or $x \approx 4.195$

Exercise 3   Solve: $x^2 + 6x - 5 = 0$

2.3 Vocabulary and Exercise

A. Fill in the blank with proper terms

1. Three methods of solving a quadratics are _____.

2. The _____ can verify if the quadratics has a solution.

3. The usage of a delta discriminant is _____.

B. Solve the following quadratics
1. $x^2 - 4 = 0$
2. $x^2 - 5x - 6 = 0$
3. $(x + 2)^2 = 0$
4. $x^2 + 3x + 2 = 0$
5. $x^2 + 2x = -3$
6. $x^2 + 6x = -9$
7. $x^2 - 8x = -16$
8. $2x^2 + 6x = -17$
9. $-x^2 + x = 0$
10. $-3x^2 + 6x = -9$

C. Graph the following quadratics and spesify the solution
1. $x^2 = y$
2. $\sqrt{y} = x$
3. $x^2 + 7x = -6$
4. $x^2 + 8x = 4$
5. $x^2 - 2x = -9$
6. $x^2 + x = -9$
7. $-x^2 + x = -9$
8. $x^2 - x + 5 = -9$
9. $x^2 + 6x = -9$
10. $3x^2 + 6x = 9$

2.4* Expansion of Quadratics and Parabola

In the previous sections, we have discussed the major properties of a quadratic equation, function, and parabola. In this section, we will be guided by an example sheet that place all properties of quadratics in practice.

Question Stem   As the graph on the right has shown, the parabola in the coordinate system xOy is modeled as $y = x^2 + bx + c$. The x-intercept of the parabola is at point A and B. The y-intercept is at point C. In graph, $OA = OC = 3$ and the vertex is at point D.

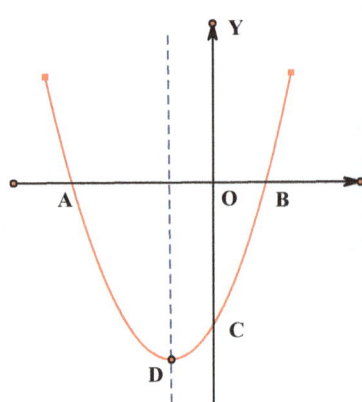

From the question, we obtained limited information. Those are:
1. The equation is $y = x^2 + bx + c$
2. The x and y intercept
3. Length of OA and OC.

However, some of the properties of a parabola which we should master.

Property 2-11   Any parabolas that can be modeled as $y = ax^2 + bx + c$ is symmetrical to the line that parallel to the y-axis and passing through the vertex.

Property 2-12   The line of symmetry of any $y = ax^2 + bx + c$ parabolas will always be equidistance from both sides.

With the information above, we can examine question 1:

Question 1   Find the equation of the parabola.

With the given information in the question, we know that the equation has only two unknown elements, b, and c. In fact, with the points demonstrated in the question stem where $OA = OC = 3$, we obtain the coordinate of point A and C as $A(-3, 0)$ and $C(0, -3)$. With these points, we obtain a system of equations with two variables: $\begin{cases} -3b+c+9=0 & (a) \\ c=-3 & (b) \end{cases}$ by substituting x and y with given coordinates. From there, we solve for b:

$$-3b + (-3) + 9 = 0$$
$$-3b = -9 + 3$$
$$-3b = -6$$
$$b = 2$$

Now, we obtain the equation for the parabola: $y = x^2 + 2x - 3$

Question 2   Connect points A, C, and D. Determine the shape of $\triangle ACD$.

This question asks the shape of the triangle. In middle school, we learned that triangle can be identified by either sides or angles. Therefore, we have:

By sides:
1. Scalene triangle: no sides are equal
2. Isosceles triangle: two of the sides are equal
3. Equilateral triangle: all three sides are equal

By angles:
1. Acute triangle: all three angles are acute angles(< 90°)
    a. Equiangular triangle: all the three angles are equal(= 60°)
2. Right triangle: one of the angles is exactly 90°
3. Obtuse triangle: one of the angles is an obtuse angle(> 90°)

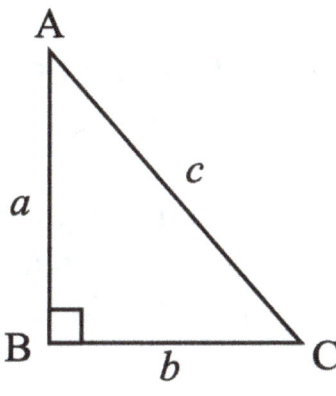

Now, as what the question stated, we connect the points, therefore having the graph above. The key in this question is to find the length of each segment. Hence, we must use the distance formula that we learned in middle school. The following is the process of obtaining the formula:

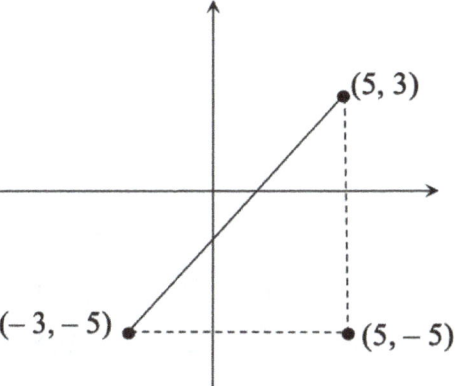

Given $\triangle ABC$ where $\angle B = 90°$. By the Pythagorean Theorem, we obtain the relationship: $a^2 + b^2 = c^2$. This characterized the relationship between length of each side. Similarly, given in a coordinate system and the coordinates of each point, we can construct a right triangle then solve for c. Thus, we obtain:

$$c^2 = a^2 + b^2$$
$$c = \sqrt{a^2 + b^2}$$

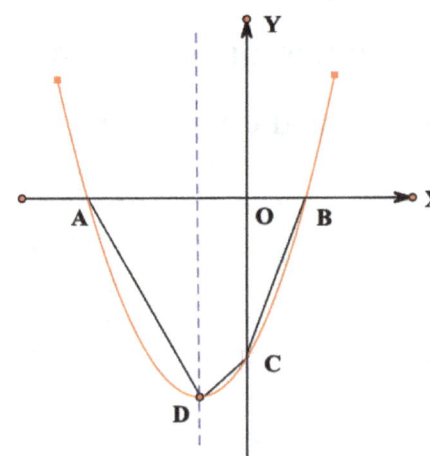

Since the length of a segment can be expressed by the difference in coordinates, by having to substitute a and b with the difference in y and difference in x, we have: $d = \sqrt{(y_2 - y_1)^2 + (x_2 - x_1)^2}$, we obtain the distance formula. As shown in the graph to the right, we have the length of b, which is eight; the length of a, which is also 8, therefore, using Pythagorean Theorem, the length of the side unknown, c, is $8\sqrt{2}$.

The proof of Pythagorean Theorem can be conducted in several different ways. One of the methods is shown below:

Given a square with the side length of $(a + b)$ units. Inside the square, another square with the side length of c units. To find the area of the small triangles, we have the following relationship: $\frac{((a+b)^2 - c^2)}{4} = \frac{ab}{2}$. From now, expand and simplify the relationship:

$$(a + b)^2 - c^2 = 2ab$$
$$a^2 + 2ab + b^2 - c^2 = 2ab$$
$$a^2 + 2ab - 2ab + b^2 = c^2$$
$$a^2 + b^2 = c^2$$

Now, to solve the question, we must find the length of segment AD. Since point D is the vertex of the polynomial, therefore using the vertex formula to obtain the coordinate of D: $(-1, -4)$.

Then find the length of segment AD: $d = \sqrt{(0+4)^2 + (-3+1)^2}$. $AD^2 = 20$. With this method, obtaining the length of the following segments: AC^2, CD^2 which are 22 and 2. Notice that $AD^2 + CD^2 = AC^2$, therefore, the shape of triangle ADC is a right triangle.

Question 3 Connect points A, B, C, and D, what is the area of the polygon?

From what observed, the polygon or the quadrilateral, is not a regular shape. To find the area of this polygon, we must draw several lines which separate the given quadrilateral into regular geometric shapes. Such lines will be drawn as:

1. Draw line passes through D which perpendicular to x-axis at point E
2. Connect points E and D
3. Connect points E and C

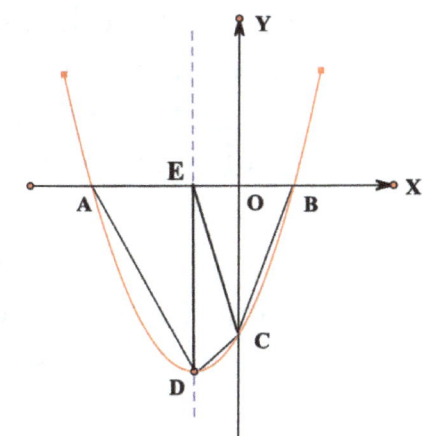

With these two lines being drawn, the quadrilateral $ABCD$ has been separated into three triangles. Since line DE is perpendicular to the x-axis, triangle AED is a right triangle which implies that segment DE is the altitude of the triangle. Similarly, segment OC is the altitude of triangle EBC.

By such relationship described above, we have the following mathematical relationship:

$$S_{ABCD} = S_{AED} + S_{EDC} + S_{EBC}$$

From there, we have the area of each triangle be: $S_{AED} = 4$, $S_{EDC} = 2$, and $S_{EBC} = 3$. Therefore, the area of the quadrilateral $S_{ABCD} = 7$.

Question 4 Will there be a point E on the y-axis that makes triangle ADE a right triangle? If so, what is the coordinate of the point? If point E does not exist, explain the reason.

From question 2, we know that all right triangles fulfill the Pythagorean Theorem. Additionally, from the equation $a^2 + b^2 = c^2$, we infer that $a + b > c$.

Therefore, the two main qualities that the answer must fulfill is defined. Then, we have the relationship.

The Pythagorean Theorem suggested that the triangle's sides' relationship. Hence, we obtain the following expression: $AE^2 + ED^2 = AD^2$. Previously(in question 2), we have found the length of AD^2 which is twenty. Thus, the following equation can be written where we set the coordinate of point E as $(0, y)$:

$$(-3-0)^2 + (0-y)^2 + (-1-0)^2 + (-4-y)^2 = 20$$

Then we expand and simplify the equation into: $y^2 + 4y + 3 = 0$. Now, we solve the quadratics by factoring it into $(y+3)(y-1) = 0$. Hence, the solution will be $y = -3$ or 1.

To verify both roots of the polynomial, we must use the rules defined previously and we found that both of the roots satisfy the rules. Therefore, we say that the coordinate of point E on the y-axis which makes triangle ADE a right triangle exists is at either $(0, -3)$ or $(0, 1)$.

Question 5 Should there be a point on the y-axis, which if connected with point A and D, creating triangle AFD that is an isosceles triangle? If not, explain the reason.

This is like question 4. Where we set the coordinate of the point F to $(0, y)$. Since the triangle is an isosceles, the length of both sides must be the same. Therefore, we have $AD = AF$. Previously(question 4), we knew that the point $(0, 1)$ satisfies the relationship by given that $AF = DF = \sqrt{10}$. Hence, we solved one of the situations where two of the sides are equal.

Then, to construct the relationship $AD = AF$, we have: $\sqrt{3^2 + y^2} = \sqrt{20}$. From here, we solve for y, which is $\pm\sqrt{11}$. Hence, the coordinate of the point F is $(0, 1)$ or $(0, \pm\sqrt{11})$.

Question 6 Construct a perpendicular line to the x-axis with the equation of $x = -1$ which intersects line AC at M, intersects the parabola at N. If the quadrilateral defined by points A, M, N, and E is a parallelogram, find the coordinate of point E.

In middle school, we learned that a parallelogram has two sets of parallel segments where each set of them have the same length.

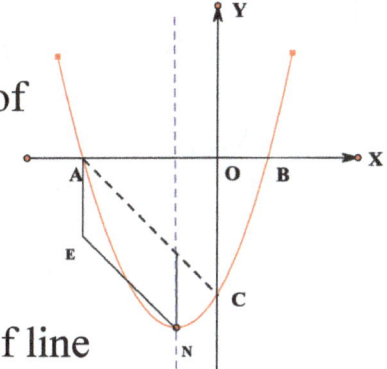

Given the parabola and the points, we can obtain the equation of line AC as $y = -x - 3$. Since ED // AC, the equation of line EN is $y = -x - 5$. Additionally, MN // AE, therefore the equation of AE is $x = -3$. Because of point E is where AE and EN intersects, we solve the equation: $y = -x - 5$ when $x = -3$. Therefore, we obtain the coordinate of the point E is $(-3, -2)$.

Question 7 Will there be a point E in the coordinate plane that makes triangle ADE a right triangle? If so, what is the coordinate of the point? If point E does not exist, explain the reason.

This question is a variant of question 4. We have the similar process but setting the coordinate of point E to (x, y). Then use the distance formula to construct the relationship:

$$(x + 3)^2 + (x + 1)^2 + y^2 + (y + 4)^2 = 20$$

Then expand and simplify the equation into the following:

$$x^2 + y^2 + 4(x + y) = -3$$

we then graph the equation, in fact, the graph is a circle. Therefore, all points on the circle will satisfy the requirements of making triangle ADE a right triangle.

LESSON SUMMARY

This section discussed the properties and applications of a parabola. The essence of the six questions above are all the combination of graphs and algebraic expressions. Such idea is critical in study of algebra. If a question cannot be understood or solved by normal algebra, it is always a great idea to graph or draw a sketch of the indicated picture, then solve the problem.

Algebra is never a pure variable or numbered field of math. In fact, majority of geometry has intersection of some sort with algebra.

Chapter 2 Summary and Review

A. Vocabulary
1. Perfect Square
2. cross-multiplication method
3. factor by grouping
4. quadratic equation
5. parabola
6. five-points rule
7. domain
8. range
9. vertex
10. inverse function
11. composite mapping
12. composite function
13. root
14. delta discriminant
15. quadratic formula
16. principle square root

B. Key Concepts
1. $a + b + c = (a + b) + c$
2. $a \cdot b \cdot c = c(ab)$
3. $a(b + c) = ab + ac$
4. $(n + n_1 + n_2 + n_3 + \ldots + n\ldots)(n + n_1 + n_2 + n_3 + \ldots + n\ldots) = n(n + n_1 + n_2 + n_3 + \ldots + n\ldots) + n_1(n + n_1 + n_2 + n_3 + \ldots + n\ldots) + \ldots + n\ldots(n + n_1 + n_2 + n_3 + \ldots + n\ldots)$
5. $a + (b - c) = a + b - c$
6. $a + (b + c) = a + b + c$
7. $a - (b + c) = a - b - c$
8. $a - (b - c) = a - b + c$
9. $a_n = a \cdot a \cdot a \cdot \ldots \cdot a$
10. $(x \pm y)^2 = x^2 \pm 2xy + y^2$
11. $(a^b)^c = a^{bc}$
12. $(a^b)(a^d) = a^{b+d}$
13. $(a^2 - b^2) = (a + b)(a - b)$
14. Cross multiplication does not apply to polynomials that exceed more than three terms
15. $ab + ac = a(b + c)$
16. $a^2 - b^2 = (a + b)(a - b)$
17. $a^3 - b^3 = (a - b)(a^2 + ab + b^2)$
18. $a^3 + b^3 = (a + b)(a^2 - ab + b^2)$
19. A quadratic equation is an equation with a squared term
20. A parabola is a U-shaped, y-axis symmetrical shaped figure.
21. Five-points rule: where we must figure out at least five points before graphing the equation.
22. A domain is the range of values for x.

23. A range is the range of values for y.

24. Write domain and range in the interval notation such that: (min, max).

25. All domain and range in interval notation will represent the max and min values for the function. Brackets indicate that the value is included while parentheses mean the value is not included.

26. The inverse of the function is first switching the function's dependent and independent variable then solve for dependent variable.

27. The notation of an inverse function is $f^{-1}(x) = ...$

28. With A_1 and A_2 are two non-empty sets. $\exists f_1$ and $\exists f_2$ as two methods which satisfy: $f_1: A_1 \to E_1$ and $f_2: E_1 \to B$. When $E_1 \subset B$, from f_1 and f_2, we can obtain a method where it will map an element $x \in A_1$ into $f_1(f_2(x)) \in B$. Then, we call the mapping from A_1 to B a composite mapping.

29. With the definition of compositional mapping, we obtain the definition of composite function as a special case where function notation used for composition mapping: with two functions: $f(x)$ and $g(x)$. With the domain of $f(x)$ as D_f and $g(x)$ as D_g, the range of $g(x)$ as R_g where $R_g \subset D_f$. then we call the function: $f(g(x))$ as a composite function.

30. $b_2 - 4ac = \Delta$.

31. $b_2 - 4ac < 0 \Leftrightarrow \dfrac{-b \pm \sqrt{b^2 - 4ac}}{2a} \notin \mathbf{R}$

32. $x = \dfrac{-b \pm \sqrt{b^2 - 4ac}}{2a}$

Chapter 2 Exercise Answer Key

2.1 Exercise

1. $(a - b)^2 = a^2 - 2ab + b^2$

2. $(x^2 + y^2)(x + y)(x - y)$
   $= x^2(x + y) + x^2(x - y) + y^2(x + y) + y^2(x - y)$
   $= x^3 + x^2y + x^3 - x^2y + xy^2 + y^3 + xy^2 - y^3$
   $= 2x^3 + 2xy^2$

3. $x^2 + 4x - 7x - 28$
   $= (x^2 + 4x) - (7x + 28)$
   $= x(x + 4) - 7(x + 4)$
   $= (x - 7)(x + 4)$

2.2 Exercise

1. $y = 2x^2$

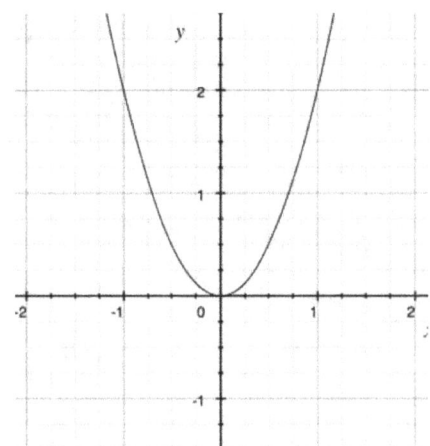

2. $f(x) = 2x^2 + 3x - 4$
   Domain: $(-\infty, +\infty)$
   Range: $[-5.125, +\infty)$

2.3 Exercise

1. $x^2 + 2x = -1$
   $x^2 + 2x + 1 = 0$
   $(x + 1)^2 = 0$
   $x = -1$

2. $(x-5)^2 + 3 = 0$

   No solution

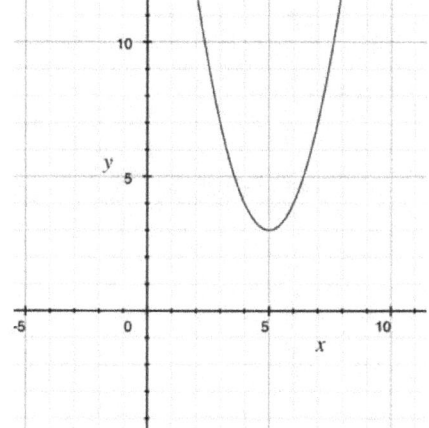

3. $x^2 + 6x - 5 = 0$
   $x - 3 \pm \sqrt{14}$

Chapter 3 System of Equations and Functions

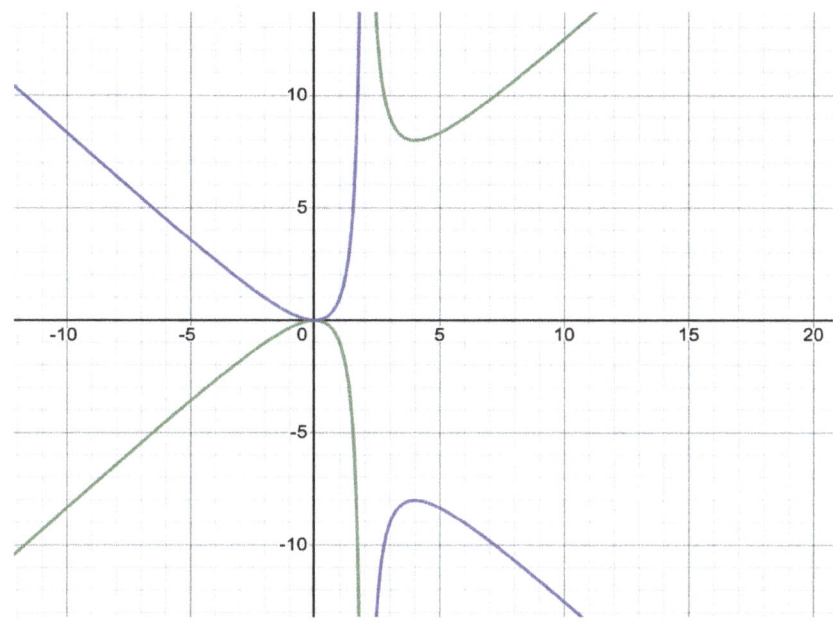

The graph of two rational equations —— $y = \dfrac{x^2}{x-2}$ and $y = -\dfrac{x^2}{x-2}$. Look that these two equations does not touch the line of $x = 2$ which is one of there asymptote.

In this chapter, we will be discussing the properties of two conic expressions and a further expansion of the systems of equations. To the field of non-linear equations and piecewise functions.

3.1 System of Equations With Three Variables

In the 1.3, we have discussed the most basic form of system of equations—with two variables. In this chapter, we will be further discussing the characteristics and types of solutions with three equations and three variables.

If an equation has three variables, we format it like: $ax + by + cz = d$. With at least three of such equations, we can find the solution set of those equations.

A common method that we use is begin with two variables, then eliminate the third variable. Therefore, creating a system of equations with two variables. Since then, solve as usual. However, notice that graphing is not usable when solving for three variables. This will result in a linear equation.

Example 1 Solve:
$$2x + 3y - z = 0 \quad (a)$$
$$x + y + z = 3 \quad (b)$$
$$2x + 6y - 2z = 0 \quad (c)$$

Key: to solve for this set of equations, we start by eliminating z, since equation (a) and equation (b) all contains z and those are inverse of each. Then solve for x, y. Substitute in an equation to solve for z.

SOLUTION

(a) + (b): $2x + 2y - z + (x + y + z) = 0 + 3$
$3x + 3y = 3$ (d)
2(a) + (c): $2(2x + 2y - z) + (4x + 6y - 2z) = 0 + 0$
$4x + 4y - 2z + 4x + 6y - 2z = 0$
$8x + 10y = 0$ (e)
$3x + 3y = 3$
$x + y = 1$
$y = 1 - x$ (f)
(f) → (e): $8x + 10(1 - x) = 0$
$8x - 10x = -10$

$$-2x = -10$$
$$x = 5 \quad (g)$$
$$(g) \rightarrow (d): \quad 15 + 3y = 0$$
$$3y = -15$$
$$y = -3 \quad (h)$$
$$(h) \rightarrow (b): \quad 5 - 3 + z = 3$$
$$z = 1$$

Exercise 1 Solve:
$$x + y - z = 3 \quad (a)$$
$$x + y + z = 3 \quad (b)$$
$$5x + 4y - 4z = 0 \quad (c)$$

Another typical way of solving a system of equations with three variables is by using **Gaussian Elimination**. Such way is to use a matrix that solve the system of equation. However, we must know what a matrix is.

Definition 3-1
A **matrix** is set of numbers, either rational or irrational, in an order as a quadrilateral.

An example of a matrix is: $\begin{bmatrix} 1 & 2 \\ 3 & 4 \end{bmatrix}$. Horizontally, the line is a row. Similarly, vertical lines are the columns. In the matrix above, row one is 1 and 2, represented as R_1. Column 1 is 1 and 3, represented by C_1. Within matrix, several operations can be done. You can change the order of any two rows but not any columns. Such as: $\begin{bmatrix} 1 & 2 \\ 3 & 4 \end{bmatrix} = \begin{bmatrix} 3 & 4 \\ 1 & 2 \end{bmatrix}$. Similarly, each row can be multiplied and added to another row. $\begin{bmatrix} 1 & 2 \\ 3 & 4 \end{bmatrix} \rightarrow \begin{bmatrix} 6 & 8 \\ 1 & 2 \end{bmatrix}$.

When using matrix or Gaussian Elimination to solve a system of equations, we first by setting a matrix into the form of equations. To accomplish that, we introduce a new form of matrix with the form of: $\begin{matrix} a & b & c & | & d \\ e & f & g & | & h \\ i & j & k & | & l \end{matrix}$. The separator separates the coefficient of the equation and the result.

To solve a system of equations with matrix, the goal is to format the matrix into the following form: $\begin{matrix} 1 & 0 & 0 & | & a \\ 0 & 1 & 0 & | & b \\ 0 & 0 & 1 & | & c \end{matrix}$. Since the numbers on the left of the separators are coefficients of variables, we obtain a system of equations in form of: ▱. In such form, we can use basic algebra to solve for all variables. However, if the matrix being solved into form such that one of the variables, does not have to be z, equals to a value, this method still applies. Consider the following example:

Example 1 Solve: $\begin{aligned} 2x + 3y - z &= 0 \quad (a) \\ x + y + z &= 3 \quad (b) \\ 6x + 7y - 3z &= 8 \quad (c) \end{aligned}$

SOLUTION

Into matrix form: $\begin{matrix} 3 & 2 & -1 & | & 0 \\ 1 & 1 & 1 & | & 3 \\ 6 & 7 & -3 & | & 8 \end{matrix}$

$\begin{matrix} 1 & 1 & 1 & | & 3 \\ 3 & 2 & -1 & | & 0 \\ 6 & 7 & -3 & | & 8 \end{matrix} \quad R_2 \to R_1$

$\begin{matrix} 1 & 1 & 1 & | & 3 \\ 0 & -2 & 0 & | & 8 \\ 6 & 7 & -3 & | & 8 \end{matrix} \quad R_2 \cdot -3 + R_3 \to R_2$

$$\begin{array}{ccc|c} 1 & 1 & 1 & 3 \\ 6 & 7 & -3 & 8 \\ 0 & -2 & 0 & 8 \end{array} \quad R_2 \to R_3$$

$$\begin{array}{ccc|c} 1 & 1 & 1 & 3 \\ 6 & 7 & -3 & 8 \\ 0 & 1 & 0 & -4 \end{array} \quad R_3 \cdot -1/2 \to R_3$$

From here, we rewrite into equation form then solve:

$$\begin{aligned} x + y + z &= 3 & \text{(a)} \\ 6x + 7y - 3z &= 8 & \text{(b)} \\ y &= -4 & \text{(c)} \end{aligned}$$

(c) → (a): $x - 4 + z = 3$
$\qquad\qquad x + z = 7 \qquad$ (d)

(c) → (b): $6x - 28 - 3z = 8$
$\qquad\qquad 6x - 3z = 36$
$\qquad\qquad 3(2x - z) = 36$
$\qquad\qquad 2x - z = 12 \qquad$ (f)

(d) + (f): $3x = 19$
$\qquad\qquad x = 19/3 \qquad$ (g)

(g) → (b): $6(19/3) - 28 - 3z = 8$
$\qquad\qquad 38 - 28 - 3z = 8$
$\qquad\qquad 10 - 3z = 8$
$\qquad\qquad -3z = -2$
$\qquad\qquad z = 2/3$

therefore, the solution of the system of equations is:
$(19/3, -4, 2/3)$

Exercise 2 Solve with matrix:
$$x + y - z = 3 \quad (a)$$
$$x + y + z = 3 \quad (b)$$
$$5x + 4y - 4z = 0 \quad (c)$$

3.1 Vocaulary and Exercise

A. Fill in the blank with proper terms

1. The basic form of a system of equations with three variables is _____.

2. A _____ is a set of ordered numbers which can perform numeral operations.

B. Identify the following system of equations has the given solution

1. $-x-5y-5z = 2$
   $4x - 5y + 4z = 19$ (1, 2, 3)
   $x + 5y - z = -20$

2. $-4x-5y-z = 18$
   $-2x - 5y - 2z = 12$. (-4, 0, -2)
   $-2x + 5y + 2z = 4$

3. $-x-5y + z = 17$
   $-5x - 5y + 5z = 5$ (3, 18, 3)
   $2x + 5y - 3z = -10$

4. $4x + 4y + z = 24$
   $2x - 4y + z = 0$ (4, 2, 0)
   $5x - 4y - 5z = 12$

C. Solve the system of equations with elimination method or substitution method

1. $x - y - 2z = -6$
   $3x + 2y = -25$
   $-4x + y - z = 12$

2. $5x + 5y + 5z = -20$
   $4x + 3y + 3z = -6$
   $-4x + 3y + 3z = 9$

3. $5x - 4y + 2z = 21$
   $-x - 5y + 6z = -24$
   $-x - 4y + 5z = -21$

4. $-5x + 3y + 6z = 4$
   $-3x + y + 5z = -5$
   $-4x + 2y + z = 13$

5. $-6x - 2y - z = -17$

$5x + y - 6z = 19$
$-4x - 6y - 6z = -20$

6. $3x - 3y + 4z = -23$
   $x + 2y - 3z = 25$
   $4x - y + z = 25$

D. Solve the system of equations with Gaussian Elimination

1. $x + y + z = 3$
   $2x + y + z = 5$
   $x + 2y - z = 4$

3.2 System of Equations With None Linear Equation

Since now, most of the equations and functions that has been discussed is ones that are **linear**. Such equations would be $f(x) = 2x + 1$. In fact, the only type of function till now that is not linear is a parabola which is self-evident that such function's graph is not a straight line. In this section, more of the none-linear equations and functions will be introduced.

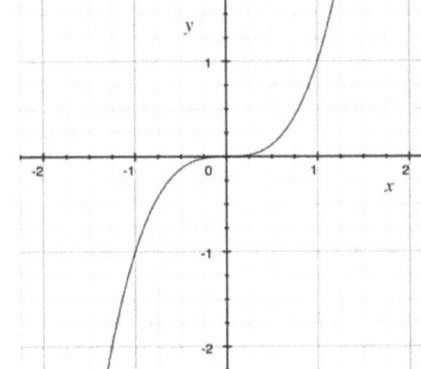

Recall from the previous chapters that a parent function is the function which no transformations has been made. Such that the parent function of a linear function is $f(x) = x$. Now try to write all parent functions that has been discussed before continuing this section.

A cubic function(equation) is where the x term is being cubed. Therefore, the parent function of that is $f(x) = x^3$. The graph of the function is shown on the right with both the range and domain of $(-\infty, +\infty)$.

From elementary school math, we know that the reverse operation of exponential operation is by taking the nth root of the number and the operation can be written into the form of a number to a fraction exponent. Such that: $2^2 = 4$, $\sqrt{4} = 2$, and $4^{1/2} = \sqrt{4} = 2$. This can be written into:

$$a^{n/m} = \sqrt[m]{a^n} \quad m \neq 0$$

therefore, we can obtain the parent function of a square root function as $f(x) = x^{1/2}$. Notice that the lower section of the graph cannot be drawn since the square root of a negative number is always undefined.

Similarly, the cube root's graph can be obtained as $f(x) = x^{1/3}$. However, the cube root of any given number is always defined. Therefore, both positive and negative number of the graph can be drawn.

Till now, all the graphs are all functions since all of them passes the vertical line test. However, the graph after this will not be a function. Such as the circle equation, ellipse, and hyperbola. However, before having such equations, we first must discuss some of the properties of the graphs.

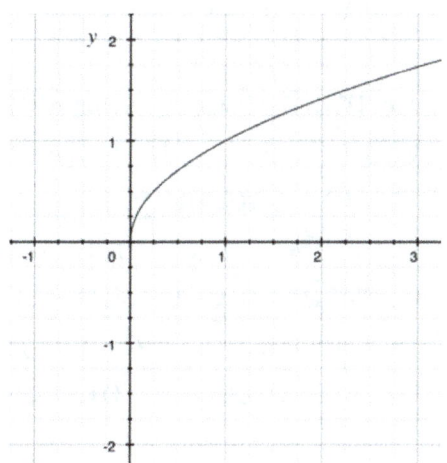

However, the graphs of the functions shown a different property. Some of the equations are more symmetrical than others. This implies the definition of an **even** or **odd function**.

An even function's graph is what called a y-axis symmetry. Such as the parabola, $f(x) = x^2$. The function is symmetrical to the y-axis compared to $f(x) = x^3$. Thus, we call functions that is symmetrical to the y-axis as an even function.

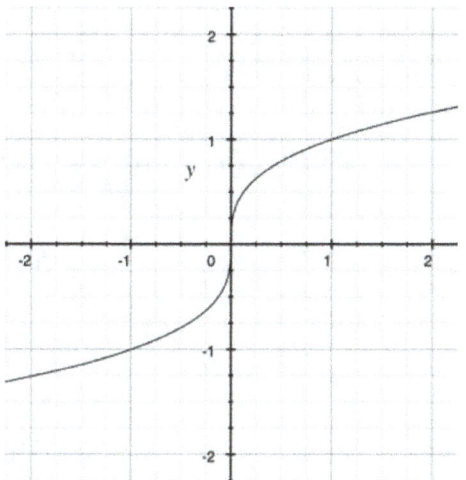

Oppositely, an odd function is more origin symmetrical. Where if the graph is turned by 180°, the graph is the same as the original. We extend this and have the following solutions and property of all parent functions.

Definition 3-2
Even function's graph has a y-axis symmetry

Definition 3-3
Odd function's graph has an origin symmetry

1) $f(x) = x$ odd function
2) $f(x) = x^2$ even function
3) $f(x) = x^3$ odd function
4) $f(x) = x^{1/2}$ neither
5) $f(x) = x^{1/3}$ odd function
6) $f(x) = a^x$ (a is a constant)   neither

7) $f(x) = 1/x \ (x \neq 0)$ neither

The method demonstrated above is to graphically determine if the function is even or odd. In fact, there is a method that will algebraically determine the function's identity. By finding $f(-x)$ and see if the function remains the same can show if the function is even. However, this does not automatically determine the function is odd if the function does not remain the same. When the function is odd, $f(-x)$ should always equals $-f(x)$. In case of the function does not meet any of the cases, the function is a neither.

Example 1   Determine $f(x) = x^3 + 2x + 1$ is an odd or even function.

SOLUTION

Given that: $f(x) = x^3 + 2x + 1$

$$f(-x) = (-x)^3 - 2x + 1$$
$$= -x^3 - 2x + 1$$

Since $f(-x) \neq -f(x)$, and $f(-x) \neq f(x)$, hence, the function is a neither.

Exercise 1   Determine if the function $f(x) = x^2 + 3x + 4$ is an odd function.

Like the ones that has linear equations, a system of equations with none-linear equations will have equations such as a circle or parabola. As mentioned, the equations do not mean to be a function. Therefore, you may encounter some equations such as a circle.

A circle equation has its basic form of $(x - h)^2 + (y - k)^2 = r^2$. By its definition, a circle is a set of points that are *equidistance* from the center.

Definition 3-4
The distance which these set of points is from the center is called the **radius**.

Definition 3-5
The point in the circle where all points are equidistance from is called the **center**.

From this definition, we imply that a circle is not a function but can be expressed using an equation. The location of the center is expressed in the equation as (h, k). It is whatever value that makes $(x - h)^2$ and $(y - k)^2$ equals to 0. The sum of these two terms is the squared radius, expressed using the letter r.

A typical question that you may encounter is to find either the radius or the center. Consider the following example.

Example 2 Determine the radius and the center of the circle that is modeled by the equation:
$(x - 3)^2 + (y - 5)^2 = 36$

SOLUTION
Given the equation of the circle as $(x - 3)^2 + (y - 5)^2 = 36$, the center is at $(3, 5)$ since it is the two values that will make the squared terms zero. The radius is 6 since $36^{1/2} = \pm 6$, because the distance cannot be a negative number, the answer must be 6.

Exercise 2 Determine the radius and center of the circle $(x + 2)^2 + (y - 3)^2 = 1$

Like a circle is what called an ellipse. From the graph, we can tell that it seems like a stretched circle. An ellipse is often obtained by having a circular cone that has been cut by a transversal plane. The surface that the cone and the plane intersect is an ellipse. In an ellipse, the sum of distance of any point on the ellipse to two of the focal points are **a constant**. Therefore, a circle is a special ellipse which two of the focal points are the same at the center. Typically, the equation which formats an ellipse is

Definition 3-6

The equation(formula) for an ellipse: $\dfrac{x^2}{a^2} + \dfrac{y^2}{b^2} = 1$

As mentioned, any ellipse will have two focal points suggested as $F = \sqrt{j^2 - n^2}$ where F is the distance which the foci from the center. j is the major axis while n is the minor axis.

Definition 3-7

The equation for the distance from foci to the center: $F = \sqrt{j^2 - n^2}$

In the equation where the ellipse is defined, the major axis will be the one terms which has the greater denominator. Such as in the ellipse: $\dfrac{x^2}{4} + \dfrac{y^2}{2} = 1$, the x-axis will be the major axis while the y-axis is the minor axis. Since there are no constants added to either x or y, the ellipse is not shifted. Therefore, the center is at (0, 0). Like the circle, the shifted equation of an ellipse is

Definition 3-8

$$\dfrac{(x-h)^2}{a^2} + \dfrac{(y-k)^2}{b^2} = 1$$

The shifting will be similar. The value which results the first or second term to be zero is the coordinate. A typical question will be the following example:

Example 3 Determine the focal points and the center of the ellipse that is modeled by the equation: $\dfrac{x^2}{9} + \dfrac{y^2}{49} = 1$

SOLUTION

With the ellipse that defined by the equation, we found that the major axis is y-axis while the x-axis is the minor axis. Hence, the focal point is $\sqrt{49 - 9}$ units from the central point at (0, 0). The graph of the ellipse is shown on the right.

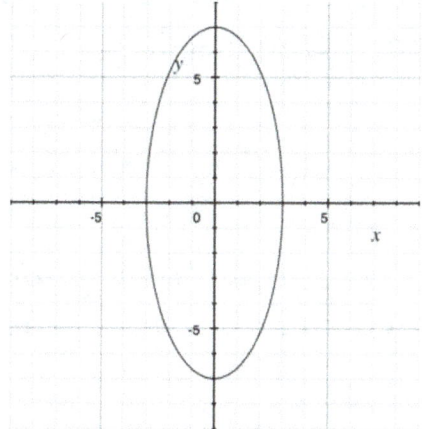

To solve a system of equations that are often none-linear, first must determine the solution type and the number of solutions that may occur in the solution set. Normally, the following cases may occur:

1) No solution: where the two graphs do not meet or intersect
2) One solution only: there will be only one solution(intersection)
3) Two or more solutions: two graphs intersects more than one times

Like what to do with two linear equations, you have one of the three methods that you may use when solving for the solutions:

1) Substitution method
2) Elimination method
3) Graphically

To solve the system of equations that has an exponential term, some of the properties which might help to solve. The first rule is if have has two exponential terms whose base is the same, their exponent must be same, this can be expressed as:

Definition 3-9
$$\forall x \in \mathbb{R}, \; x^a = x^b \rightarrow a = b$$

In fact, majority of the question encountered in this chapter does not need this rule, however, when solving logarithmic or exponentials, this rule will be frequently used.

Example 4 Solve the following: $(x+y)^2 = 0$ (a), $y - 2x = 2$ (b).

Key: start by solving for y in the second equation, then substituting y in the first equation, solve from there.

SOLUTION
with the system of equations:
$$(x+y)^2 = 0 \quad (a)$$
$$y - 2x = 2 \quad (b)$$

(b) $\quad y = 2x + 2 \quad\quad\quad$ (c)

(c) → (a) $x^2 + 2xy + y^2 = 0$
$x^2 + 2x(2x+2) + (2x+2)^2 = 0$
$x^2 + 4x^2 + 4x + 4x^2 + 8x + 4 = 0$
$9x^2 + 12x + 4 = 0 \quad\quad$ (d)

Then solve for x and y. We obtain: $(-0.85, 0.301)$ and $(-0.262, 1.477)$

3.2 Vocabulary and Exercise

A. Fill in the blank with proper terms

1. Equations that follows a contant rate of change is called a _____ equation(function).

2. All graphs of any even functions has a _____.

3. All graphs of any odd functions has an _____.

4. A circle has the definition of: a set of points that are _____ from the _____.

5. Radius is the _____ of _____ to the center.

6. The point which all points are _____ from is called the _____.

7. The equation for an ellipse is _____.

8. The distance from the foci to the center on an ellipse is modeled by equation _____.

9. A shifted ellipse can be modeled by _____.

10. $\forall x \in \mathbb{R},\ x^a = x^b \rightarrow$ _____.

B. **Determine the center, radius and foci of the following, you may graph the equation as assist**
1. $x^2 + y^2 = 36$
2. $(x + 4)^2 + y^2 = 9$
3. $(x - 3)^2 + (y + 1)^2 = 4$
4. $2(x - 4)^2 + 2y^2 = 8$
5. $x^2 = -y^2 + 144$
6. $\frac{(x-2)^2}{4^2} + \frac{(y-4)^2}{16^2} = 1$
7. $\frac{x^2}{16^2} + \frac{(y-8)^2}{7^2} = 1$
8. $\frac{x^2}{3^2} + \frac{y^2}{16} = 1$
9. $4x^2 + 9y^2 = 36$
10. $9x^2 + 7y^2 = 63$

C. **Fill in the blank with proper parent function, then graph the parent function**
1. The parent function of a parabola is _____.
2. $f(x) = x^3$ is the parent function of _____.
3. $f(x) = x^{1/2}$ is equivalent to _____.
4. The function $f(x) = x^4$ is a _____ (even / odd) function.
5. Name three different parent functions that are odd 1. _____. 2. _____. 3. _____.

D. **Solve the following**
1. $y = x^2 - 4$
   $y = -x - 2$

2. $y = -x^2 - 2x + 14$
   $y = x^2 - 4x - 10$

3. $x^2 + y^2 = 4$
   $y^2 - x = 4$

4. $x^2 = 2y + 10$
   $3x - y = 9$

5. $2^x = 2^y$
   $x + y = 7$

3.3 Piecewise Functions and Discontinuity

For the past, all function which we had discuessed are all **continual** at majorty. Therefore, what is a **discontinuity**? This critical question leads to the defintion of limitation.

Definition 3-10

Given a sequence of real numbers, denoted as $\{a_n\}$, with a constant β, for any postitive number k, there always be a postive integer K, which, let $n > K$. That satisfies the inequality $|a_n - \beta| < k$, therefore suggesting that β is the limit of sequence $\{a_n\}$, noted as $\lim\limits_{n \to K} a_n = \beta$.

In simpler terms, with a given function such as $f(x) = \dfrac{x^2 - 1}{x - 1}$, we can appened all is dependent variables to a sequence. However, we found that as the number we choose approches 1, the output of the function also approches to a contant. However, the function is not defined at the point of 1. Therefore we suggested that the constant is the limit of this function. To better understand this concept, try think the following way:

You are given an apple pie. However, you noughty brother decided to only eats the fillin which left you something simlar to a tortilla. In such case, this can be repersented as $\lim\limits_{fillin \to 0} pie = tortilla$.

The notation of limitation in Definition 3-10 can be read as "The limitation of sequence a as n approaches K is β".

To find the limit of a function, there are methods which we can follow. First is to make a table which represents the limit of the function. In such methods, you fillout a table that is similar to the x-y table when graphing, scrutenize the table and conclude with the limit. For the majority of linear functions, the limit is the output of the function at the maximum point. Consider the following example:

Example 1   Find the limit of the function $f(x) = 5x + 3$ as it approaches 5.

Key: If to graph the function, we found that the function has not discontinuity at the point of 5, therefore, simply evaluate the function at 5.

SOLUTION
Given the information, we have the following notation: $\lim_{x \to 5} 5x + 3 = ?$

$$5x + 3 \Big|_{x=5} = 5 \times 5 + 3 = 28$$

Therefore, the limit of $f(x)$ as x approaches 5 is 28.

Exercise 1   Find the limit $f(x) = 7x$ as it approaches 7.

However, some limits cannot easily be evaluated, such as the one shown above. $f(x) = \dfrac{x^2 - 1}{x - 1}$. If to evaluate $\dfrac{x^2 - 1}{x - 1}\Big|_{x=1}$, we found that the function is undefined, as mentioned above. Such equations are called an indeterminate form.

Definition 3-11
If two functions, $f(x)$ and $g(x)$, both approaches 0 or positive infinity, the fraction which these two form is called an indeterminate form.

When finding such limites of the function, we have to use the L' Hospital Rule. This rules was proposed by Guillaume de l'Hôpital in 1696. The rules suggested that:

Definition 3-12

Given an indeterminate form, $\frac{f(x)}{g(x)}$, the limitation of such fraction, $\lim\limits_{x \to k} \frac{f(x)}{g(x)}$ $k \in R$. If $\lim\limits_{x \to k} f(x) = 0$ $k \in R$ and $\lim\limits_{x \to k} g(x) = 0$ $k \in R$ and derivable at a given point a, which, $g'(x) \neq 0$. Therefore having

$$\lim_{x \to k} \frac{f(x)}{g(x)} = \lim_{x \to k} \frac{f'(x)}{g'(x)} \quad k \in R.$$

The calculation of L' Hospital Rule involves the calculation of a function's derivative. Recall that given any linear equation, the slope is the constant rate of change of the function. Such example will be $f(x) = 2x - 1$. The slope is 2. However, a derivative is the instant rate of change on a non-linear function such as $k(x) = x^2$. The definition is shown below:

Definition 3-12

Having a tangent of any line(curve). The rate of change of the independent variable x at any given point p, can be represented as an increasement, denoted as Δx. From the equation of slope, $m = \frac{y_2 - y_1}{x_2 - x_1}$, therefore obtain the slope of the tangent as $m_{tangent} = \frac{f(x + \Delta x) - f(x)}{\Delta x}$.

However, when the increasement of the independent variable, Δx, approches 0, the fraction will be undefined. In fact, the closer which we approches to 0, the slope of tangent will reach to a constant. Therefore, the derivative of the function can be defined and denoted as:

Definition 3-13

The rate of change of the tangent of the curve, denoted as $f'(x)$, is as the increasement of independent variable approches 0. Therefore, having the

increasement, h, we have the basic equation of derivative:
$$f'(x) = \lim_{h \to 0} \frac{f(x+h) - f(x)}{h}.$$

Example 2 Find the derivative of function $f(x) = x^2$ at 5.
Key: Use the equation of derivate to find the derivative of x^2

SOLUTION
Have the function $f(x) = x^2$, obtaining:

$$f'(x) = \lim_{h \to 0} \frac{(x+h)^2 - x^2}{h}$$
$$= \lim_{h \to 0} \frac{x^2 + 2xh + h^2 - x^2}{h}$$
$$= \lim_{h \to 0} \frac{2xh + h^2}{h}$$
$$= \lim_{h \to 0} \frac{h(2x + h)}{h}$$
$$= \lim_{h \to 0} 2x + h$$
$$= 2x + h \Big|_{h=0} = 2x$$
$$= 2x \Big|_{x=5} = 10$$

Exercise 2 Find the derivative of $f(x) = x^3 + 4$ at 7.

Therefore, we conclude with the first property of derivative: the derivative of exponential functions as

Property 2-12 The derivative of any expoential function
$a^k = ka^{k-1} \quad k \neq 0$.

Example 3 Find the derivative of a constant C at any point of coordinate.

SOLUTION

$$f'(x) = \lim_{h \to 0} \frac{(x+h)^2 - x^2}{h}$$
$$= \lim_{h \to 0} \frac{C - C}{h}$$
$$= \lim_{h \to 0} 0$$
$$= \lim_{h \to 0} 0 = 0$$

From this example, we conclude that the derivative of any constant is always 0.

Property 2-13 The derivative of any constant $C' = 0$.

Similarly, we can obtain the main rules of derivatives:

1. $a^k = ka^{k-1} \quad k \neq 0$
2. $C' = 0$
3. $(f(x) + g(x))' = f'(x) + g'(x)$
4. $(Cf(x))' = Cf'(x)$
5. $(f(x)g(x)) = g(x)f'(x) + f(x)g'(x)$

6. $\left(\dfrac{f(x)}{g(x)}\right)' = \dfrac{f'(x)g(x) - g'(x)f(x)}{g^2(x)}$

Another type of discontinuity is created by piecewise functions. The piecewise function has the basic form of $f(x) = \begin{cases} f_1(x) & if... \\ f_2(x) & if... \end{cases}$. To evaluate the function at a certain point, we have first determine which function to be use. Consider the following example:

Example 4 Evaluate $f(x) = \begin{cases} 2x+3 & if\ x \leq 7 \\ x^2+16 & if\ x > 7 \end{cases}$ when $x = 10$ and $x = 7$.

SOLUTION

$$\begin{cases} 2x+3 & if\ x \leq 7 \\ x^2+16 & if\ x > 7 \end{cases}$$

When $x = 10$, $x > 7$, then $f(10) = 10^2 + 16 = 116$

When $x = 7$, $x \leq 7$, then $f(7) = 2(7) + 3 = 17$

Exercise 3 Evaluate $f(x) = \begin{cases} 5 & if\ x \leq 1 \\ 2x & if\ x > 1 \end{cases}$ when $x = 15$ and $x = -1$

When graphing a piecewise function, it is important to notice which function at what domain should be used. When the condition has a greater(lesser) sign, there should be a hole on the graph. Otherwise there should be a solid circle on the graph. Consider the following example:

Example 5 Graph the piecewise function:
$f(x) = \begin{cases} 3 & if\ x \leq 0 \\ 4 & if\ x > 0 \end{cases}$

SOLUTION

The piecewise function are two horizontal lines, having a whole on the y-axis. Hence the graph is shown as:

Exercise 4 Graph the piecewise function $f(x) = \begin{cases} x+1 & \text{if } x \leq 0 \\ 4x & \text{if } x > 0 \end{cases}$

3.3 Vocabulary and Exercise

A. Fill in the blank with proper terms

1. As the sequence approches a certain constant, the constant is called the _____.

2. The limit of a sequence is denoted as _____.

3. $\lim\limits_{x \to a} f(x)$ is read as _____.

4. The rate of change at a certain point on graph is called the _____.

5. The function which modeled the relationship is called _____.

6. Derivative has the equation of _____.

B. Find the following limits

1. $\lim\limits_{x \to 6} 2x + 3$
2. $\lim\limits_{x \to -1} 3$
3. $\lim\limits_{x \to 6} \dfrac{x}{x-6}$
4. $\lim\limits_{x \to 1} \dfrac{x^2 - 1}{2x}$
5. $\lim\limits_{x \to 0} \dfrac{1}{x}$

C. Find the derivative of the functions(expressions)

1. $y = 2x$
2. $y = 3$
4. $y = x^5$
5. $y = \dfrac{x}{3x}$
6. $y = \sqrt{x}$
7. $y = \sqrt{x} + 8x + 2$
8. $y = 3x^4 - 5$
9. $y = ax \quad a \in R$
10*. $y = \sin x$

D. Evaluate then graph the piecewise function

1. $f(x) = \begin{cases} x+1 & \text{if } x \leq 0 \\ x^2 & \text{if } x > 0 \end{cases} \quad x = 4$

2. $f(x) = \begin{cases} -x & \text{if } x \leq 5 \\ x & \text{if } x > 5 \end{cases} \quad x = -1$

3. $f(x) = \begin{cases} \sqrt{x} & \text{if } x \leq 5 \\ -x+3 & \text{if } x > 5 \end{cases} \quad x = 4$

4. $f(x) = \begin{cases} 1 & \text{if } x \leq 0 \\ 0 & \text{if } x > 0 \end{cases}$ $\quad x = -2$

5. $f(x) = \begin{cases} 0 & \text{if } x \notin R \\ 3x+14x^2 & \text{if } x \in R \end{cases}$ $\quad x = i$

(Do not graph)

Chapter 3 Summary and Review

A. Vocaulary
1. System of Equations with 3 Variables
2. Gaussian Elimination
3. Matrix
4. Even function
5. Odd function
6. Ellipse
7. Circle
8. Radius
9. Center
10. Foci
11. Focal points
12. Discontinuity
13. Limit
14. Derivative
15. Piecewise Functions
16. Instant Rate of Change
17. L' Hospital Rule of Limitation

B. Key Concept
1. If an equation has three variables, we format it like: $ax + by + cz = d$
2. A matrix is set of numbers, either rational or irrational, in an order as a quadrilateral.
3. Another typical way of solving a system of equations with three variables is by using Gaussian Elimination.
4. $f(x) = x$
5. $f(x) = x^3$
6. $a^{\frac{n}{m}} = \sqrt[m]{a^n}$ $m \neq 0$
7. An even function's graph is what called a y-axis symmetry.
8. An odd function is more origin symmetrical.
9. Even function $f(-x) = f(x)$
10. Odd function $f(-x) = -f(x)$
11. By finding $f(-x)$ and see if the function remains the same can show if the function is even.
12. This does not automatically determine the function is odd if the function does not remain the same.
13. In case of the function does not meet any of the cases, the function is a neither.
14. Circle's equation: $(x - h)^2 + (y - k)^2 = r^2$
15. A circle is a set of points that are *equidistance* from the center. The distance which these set of points is from the center is called the radius. The point in the circle where all points are equidistance from is called the center.

16. The location of the center is expressed in the equation as (h, k).

17. In an ellipse, the sum of distance of any point on the ellipse to two of the focal points are a constant.

18. A circle is a special ellipse.

19. $F = \sqrt{j^2 - n^2}$

20. $\dfrac{x^2}{a^2} + \dfrac{y^2}{b^2} = 1$

21. $\forall x \in \mathbb{R},\ x^a = x^b \rightarrow a = b$

22. Given a sequence of real numbers, denoted as $\{a_n\}$, with a constant β, for any postitive number k, there always be a postive integer K, which, let $n > K$. That satisfies the inequality $|a_n - \beta| < k$, therefore suggesting that β is the limit of sequence $\{a_n\}$, noted as $\lim\limits_{n \to K} a_n = \beta$.

23. If two functions, $f(x)$ and $g(x)$, both approches 0 or positive infinity, the fraction which these two form is called an indeterminate form.

24. Given an indeterminate form, $\dfrac{f(x)}{g(x)}$, the limitation of such fraction, $\lim\limits_{x \to k} \dfrac{f(x)}{g(x)}\ \ k \in R.$ If $\lim\limits_{x \to k} f(x) = 0\ \ k \in R$ and $\lim\limits_{x \to k} g(x) = 0\ \ k \in R$ and derivable at a given point a, which, $g'(x) \neq 0$. Therefore having
$$\lim\limits_{x \to k} \dfrac{f(x)}{g(x)} = \lim\limits_{x \to k} \dfrac{f'(x)}{g'(x)} \quad k \in R.$$

25. Having a tangent of any line(curve). The rate of change of the independent variable x at any given point p, can be represented as an increasement, denoted as Δx. From the equation of slope, $m = \dfrac{y_2 - y_1}{x_2 - x_1}$, therefore obtain the slope of the tangent as
$$m_{tangent} = \dfrac{f(x + \Delta x) - f(x)}{\Delta x}.$$

26. The rate of change of the tangent of the curve, denoted as $f'(x)$, is as the increasement of independent variable approches 0. Therefore, having the increasement, h, we have the basic equation of derivative:
$$f'(x) = \lim\limits_{h \to 0} \dfrac{f(x + h) - f(x)}{h}.$$

27. $a^k = ka^{k-1} \quad k \neq 0$

28. $C' = 0$

29. $(f(x) + g(x))' = f'(x) + g'(x)$

30. $(Cf(x))' = Cf'(x)$

31. $(f(x)g(x)) = g(x)f'(x) + f(x)g'(x)$

32. $$\left(\frac{f(x)}{g(x)}\right)' = \frac{f'(x)g(x) - g'(x)f(x)}{g^2(x)}$$

Chapter 3 Exercise Answer Key

3.1 Exercise
1. (– 12, 15, 0)

2. Same as above.

3.2 Exercise
1. $f(x) = x^2 + 3x + 4$
   $f(-x) = (-x)^2 - 3x + 4$
   $f(-x) = x^2 - 3x + 4$
   The function is neither odd nor even.

2. $(x + 2)^2 + (y - 3)^2 = 1$
   Center: (– 2, 3), Radius: 1

3.3 Exercise
1. 49
2. 147
3. When $x = 15, f(15) = 30$; When $x = -1, f(-1) = 5$
4.

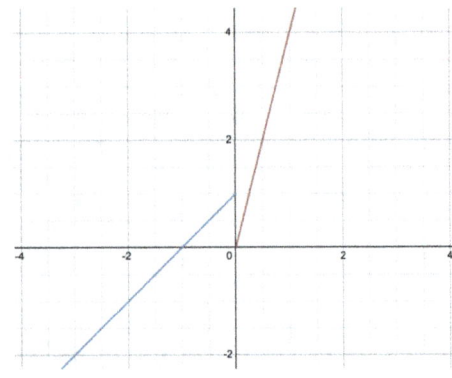